Global Responses to AIDS

Global Responses to AIDS

Science in Emergency

Cristiana Bastos

Indiana University Press

Bloomington and Indianapolis

This book is a publication of

Indiana University Press
601 North Morton Street
Bloomington, Indiana 47404-3797 USA

www.indiana.edu/~iupress
Telephone orders 800-842-6796
Fax orders 812-855-7931
Orders by e-mail iuporder@indiana.edu

The paper used in this publication meets the minimum requirements of American National Standard for Information Sciences—Permanence of Paper for Printed Library Materials, ANSI Z39.48-1984.

Manufactured in the United States of America

Library of Congress Cataloging-in-Publication Data

Bastos, Cristiana.
Global responses to AIDS : science in emergency / Cristiana Bastos.
p. cm.
Includes bibliographical references and index.
ISBN 0-253-33590-6 (cl. : alk. paper). —
ISBN 0-253-21335-5 (pa. : alk. paper)
1. AIDS (Disease)—Social aspects. 2. AIDS (Disease)—Brazil—Rio de Janeiro. I. Title.
RA644.A25B358 1999
362.1'969792—dc21 99-32161

1 2 3 4 5 04 03 02 01 00 99

Contents

Acknowledgments

This book is the result of an extended experience that began with a Ph.D. in Anthropology and continued with several types of fieldwork and the processing of their impact (1989–96). The experience was spread throughout several places and continents: New York City; Rio de Janeiro and other Brazilian sites; the yearly gatherings for the AIDS International Conference in San Francisco, Florence, Amsterdam, and Berlin; and, in a later phase, at electronic sites. Rather than having the mythical, pristine ethnography of a small community, the ethnographic method was pushed to the limits, using some of the features of contemporary culture—the transnational communities and networks, the primacy of information, the compression of space and time, and the increasing social gaps in the world. These subjects are at the center of current academic theory, and the methodology to research them is still in the making. That fact adds a feeling of unsettledness when processing the experience into knowledge. In my research there were additional unsettling factors: as much as I was focused on the production of knowledge and the work of scientists under the constraints of an asymmetrical global network, I was immersed in a world that meant, for many, negotiating one's life on a daily basis, scrutinizing research news on treatments and basic science, and inventing collective and individual coping strategies. Many of my field friends died before the triple cocktail came to help many others. To both groups go my first acknowledgments.

I also thank the many people who helped me in different institutions—scientists, physicians, health professionals, activists, colleagues, advisors, funders. The Instituto de Ciências Sociais in Lisbon and the Graduate Center at the City University of New York provided me with extensive support and guidance, a research environment, and the motivation to work and to face great challenges. The INVOTAN-JNICT grant program, the Wenner-Gren Foundation for Anthropological Research, Fulbright, Delta Kappa Gamma, CUNY Research Foundation, and ICS provided me with the financial support that made possible the several phases of research. The Graduate Program in Anthropology at the Museu Nacional, Federal University of Rio de Janeiro (UFRJ), the Instituto de Medicina Social, of the State University of Rio de Janeiro (UERJ), and the Associação Brasileira Interdisciplinar de AIDS (ABIA) provided me with an exceptionally intellectual and friendly atmosphere. The AIDS unit at University Hospital at UFRJ and the research units on tropical medicine, infectious diseases, immunology, public health, and history of medicine at Fundação Oswaldo Cruz, gave me the ideal field sites to pursue my questions. Further hospitality came from the PWA group Pela VIDDA, from the Ministry of Health in Brasília, from the Global Programme on AIDS at the World Health Organization in Geneva, and from the many international institutions and networks that I came across. Working as a freelance reporter for the media gave me privileged access to the information and people that circulated during the International Conferences on AIDS.

I owe special thanks to the people I met in those institutions and from whom I learned what I processed in the text: in Rio, Sergio Carrara, Richard Parker, Kenneth Camargo, Dina Czeresnia, Jane Galvão, Veriano Terto, Carmen Dora Guimarães, Herbert Daniel, Claudio Mesquita, Silvia Ramos, Walter Almeida, José Stalin Pedrosa, Christina Vallinotto, Cristina Câmara, Simone Monteiro, Lys Portella, Cristina Alvim, José Carlos Lopes, Antonio Carlos de Souza Lima, Gilberto Velho, Ana Dao, Claudia Moraes, Luiz Fernando Duarte, Andrea Loyola, Madel Luz, Luiz Antonio Castro Santos, Marilena Correa, Letícia Legay, Regina S. Barbosa, Maurício Peres, Murilo Peixoto, Marlene Zornitta, Gisela Câmara, Mauricio Tostes, Betina Durovni, Andrea Sereno, Mauro Schechter, Celso Ramos Filho, Paulo Feijó, Valdilea Veloso, Cydia, Helena, Graça, Suzi, Peixoto, Márcia Rachid, Beatriz Grijnztein, Frits Sutmoller, Albanita Vianna, Euclides Castilho, Sérgio Coutinho, Mariza Morgado, Claudio Ribeiro, Eloy Garcia, Claude Pirmez, Ricardo Ribeiro dos Santos, Haity Mussachê, Wilson Savino, George dos Reis, Alvaro Matida, Regina Guedes, Claudia Damasceno, Paulo Chagastelles Sabroza, Rosa Salerno Soares, Ana Filgueiras, Patrick Larvie, Paulo Longo, Márcia Suzart; in São Paulo, Nelson Solano, Maria Eugenia Fernandes, Jorge Beloqui; in Bahia, Bernardo Galvão-Castro, Naomar Almeida, Luiz Mott; in Belo Horizonte, Nelson Vaz, Dirceu Greco; in Brasília, Lair Guerra, Pedro Chequer, Paulo Teixeira, Eduardo Cortes; in New York, the faculty and students at the Graduate Center—Shirley Lindenbaum, Vincent Crapanzano, Jane Schneider, Louise Lennihan, Leith Mullings, Ulf Hannerz, Ara Wilson, Carmen Medeiros, Tom Burgess, Lawrence Hammar, Yvonne Lassale, Maureen O'Daugherty, Alfredo Gonzalez, Helio Belik, Ligia Simonian, Patricia Tovar, Treni Jungens, and many others, plus

Carlos Gibson, Mike Dyckman, Trina Marx, Kate Wilson, and the many non-academic friends from whom I learned the most; in Lisbon, to colleagues at ICS, in particular João Pina Cabral and Manuel Villaverde Cabral, I owe the confidence to pursue what at times seemed overwhelming. Special thanks to those who commented on the ideas in the manuscript—besides the ones already mentioned—Meurig Horton, Peter Aggleton, Sarah Franklin, Susan DiGiacomo, anonymous reviewers, and my editors at Indiana University Press, Jane Lyle and Joan Catapano.

Introduction

After the announcement in 1996 of the good clinical results of a new combination of anti-HIV drugs, many people rejoiced. For them it was the long-awaited news of a possible cure for AIDS. Activists, patients, doctors, scientists, politicians, and many others fought hard for that moment. Did it mean that the epidemic was over, that people could retreat from their sheltered, suspended lives, from the feeling of having lost the strength and stamina to go on? Had it restored the public's belief in doctors' ability to treat and cure, after years of discrediting them? And did it mean that people no longer needed to fear sex and exposure to blood as potential death traps? Or that the Third World would no longer lose a few extra million lives?

While many rejoiced, others were cautious about the meaning of the news. The new "triple cocktail" could turn out to be another AZT—a drug that nine years before had triggered so many expectations only to show, in the end, disappointingly limited results. The new drugs could also offer no more than a temporary respite along a very long and arduous journey to a cure. After all, the triple cocktail introduced the recently tested protease inhibitors to the drug supply for AIDS. Could the use of these agents lead to the development of multi-resistant strands of HIV capable of being even more lethal than the ones that are present now?

Moreover, being extremely expensive, the new drugs will not be available for everyone. They may change the face of the pandemic from that of a *worldwide* problem to that of a *Third World* problem—a disease which, like so many others, has a cure that only the affluent can afford. It was precisely the lack of a cure for AIDS—which brought together rich and poor, developed and underdeveloped, centers and peripheries for a decade and a half—that had made this epidemic so unique. The disease had affected those who were beyond the reach of typical epidemics, i.e., the affluent and politically empowered. And partly for that reason, they were also able to respond with action and intervention, lobbying and creativity, introducing a change in the traditional ways of responding to disease and promoting biomedical research.

The new epidemic caught urban America by surprise during the flamboyant and prosperous 1980s. AIDS challenged the power of medicine, pushing the limits of medical knowledge and demonstrating how illusory their history of irreversible victories against infectious disease really was. Death, defeat, and collective mourning seeped into the daily life of communities that had little experience with them. For a long time, no cure was in sight.

This was not just another medical crisis. It was also a crisis in knowledge and politics. In response to the epidemic, some of the people whose lives were at risk took action into their own hands, and a new social movement started. It contested, lobbied, and eventually worked together with the physicians and scientists. From that encounter, new forms of knowledge were born.

Different social partners produced new knowledge: That is the subject of this book. Arguing for the creative and transformative potential of those encounters, my analysis accounts for two types of interactions: the encounters between the AIDS social movement and the medical establishment, and the institutionally sponsored transnational partnerships between "central" and "peripheral" world forces. The social production of biomedical knowledge will be studied within the context of the heterogeneities and tensions in the contemporary global society, as they are revealed by the AIDS crisis.

Such a project can hardly be contained in a linear narrative. Every issue unfolds within many others, and each fold is overloaded with information—contradictory, unbalanced, and unfinished. This is not a "strong narrative," but a multi-layered, composite approach that involves multi-sited ethnographic research in different spheres of the collective responses to AIDS: AIDS activism in New York City and the United States more generally, the World Health Organization, the nongovernmental organizations (NGOs) in South America, and the clinical settings and research centers for infectious disease in Brazil.

The text begins with an overview of state-of-the-art biomedical knowledge and social commentary about it. This is followed by an analysis of the social movement generated by the inability of the medical establishment to respond to the AIDS crisis efficiently. It then covers the growing awareness of the global dimensions of the epidemic and the efforts of international agencies to create the materiel necessary to launch an effective global response. The text then examines a field setting where the contradictions and asymmetries of the contemporary world are combined in a single place. That place corresponds to Rio de Janeiro, involving health care centers, community organizations, and AIDS agencies. Finally, the text attempts to evaluate the extent to which global responses to AIDS lead to the creation of interactive structures with an influence in the production of transformative knowledge.

How was this research accomplished? Not through the traditional eighteen-month residency in a field community, but through a somewhat updated version of fieldwork, one adapted to the subject: multi-sited, scattered, transnational, moving along a subject of multiple meanings and in constant transformation. It started after I moved to New York City in 1987 for academic reasons, and became a witness of the health crisis as well as of the community response to it. I was not formally involved with New York AIDS agencies, as a researcher, volunteer, or agency staff member, so my knowledge of the New York movement is presented in this book not as "data" or "insider's knowledge," but rather as the knowledge of a close witness. Sources of information included newspapers, handouts, leaflets, videos, films, public speeches, rallies, word of mouth, and participation in ACT UP meetings. There was also analytical material produced within that movement. At the time, it attracted a number of writers, social scientists, and critical theorists, and from them come many of the references and inspiring sources in Chapter 2. A few journalistic accounts of the epidemic and responses to it in America complement those sources. Contact with this large body of printed knowledge eroded the boundaries between direct observation and a more academic type of learning. The "observation/learning mode" would come back time and again during the project, through formal field work in Brazil and later through the processing of "research data."

The global setting, which the reader is constantly reminded of in the text, is treated separately in Chapter 3. A more "ethereal" field site than New York City, the global sphere is also more than merely a device of rhetoric. It is somehow a political site with a material basis in international funding and the cognitive dimension of global awareness about AIDS. Its physicality was present in full color in the ritual moments of international conferences, where all the partners and problems gathered; it was maintained through the year in the "colder" structures and routine of international agencies.

The international agencies linked to development, external aid, and the United Nations, oriented since World War II toward understanding and alleviating the painful gaps in the world, took on the double task of defining the global dimensions of AIDS and of framing the regional and macro differences in the experiences of the pandemic. It was their action-oriented knowledge, rather than the callous work of anthropologists theorizing about world heterogeneities, that resulted in the identification of those experiences. Responding rapidly to the prospect of a global health crisis, WHO created the Global Programme on AIDS (GPA). In 1988, it summarized the diversity of forms of the epidemic within a three-pattern model; from then on, there would be "pattern I," "pattern II," and "pattern III" countries or regions. In popular knowledge and in the media, this model corresponded to the contrast between a "developed countries pattern," centered on homosexual and drug-related transmission, and a "developing countries pattern," centered on heterosexual transmission, plus a large pattern-block with no reported cases and yet the potential for an epidemic explosion. The key concepts for defining those patterns, and their popular/mediatic interpretation, were based on individual behavior and drawn from the cultural categories of the urban centers where the epidemic had first been diagnosed.

This fact raised a problem only too familiar to those who analyze and deconstruct the discursive formations of science: The very description of the worldwide epidemic that was to serve as an instrument for further analysis was itself a social construction that assumed that local categories were universal. In 1988 and 1989, I discussed this problem in a paper on the cultural construction of epidemiology. The need to substantiate the points transformed the paper into a research proposal which led me to Brazil, where the search for a definition of AIDS epidemiology was being conducted on a battleground of dissent, criticism, and negotiation, as expressed in the works of social scientists and political writers (see Chapters 3 and 4).

In two preliminary field trips to Brazil in 1989 and 1990, I contacted the local institutions where this critical stance was more visible: ABIA (the Brazilian Interdisciplinary AIDS Association), a Rio-based NGO created in late 1986; and the Social Medicine Institute, a department of the State University of Rio de Janeiro, where the focus is on the convergence of the fields of health and social sciences. I also contacted other local academic departments, activist organizations, and health services to plan my fieldwork. The fieldwork, which lasted from December 1990 to May 1992, was urban and was scattered over a wide area, to include several institutions in Rio de Janeiro and short visits to São Paulo, Salvador, Belo Horizonte, Brasília, and other cities. The inclusion of such a variety of "field sites" was necessary to prevent making overly broad generalizations about AIDS in Brazil. However, my conclusions result from having been based in Rio.

By increasing the diversity of social actors contacted for this study, I was able to go beyond discussion of the cultural construction of epidemiology and examine the broader aspects of medical knowledge that had interested me in the first place. I finally gave priority to the analysis of basic science, while taking clinical medicine into account. A similar focus was given by an important component of AIDS activism in New York City to the examination of immunology and the biology of infectious disease. At that time, this was believed to be the area where a significant leap in knowledge might radically affect the face of AIDS. After examining the creative interactions between social partners of different backgrounds (such as AIDS activists and scientists) in urban North America, I looked for potentially creative interactions between Brazilian medical researchers and mainstream international researchers. Concentrating initially on a university hospital AIDS unit that combined medical care and research (UFRJ), I moved later to a biomedical research center that covered other areas of infectious disease and immunology research (FIOCRUZ).

Hospitals, NGOs, government offices, health centers, and research institutes provided the reference sites for my definition of the terrain. Leisurely activities filled in my understanding of the local issues. Informal conversations were combined with formal interviews. Institutional moments of definition and evaluation, such as scientific and organizational meetings, provided a structured self-portrait of the process. Participation in the daily life of the agencies and health care centers permitted the anthropological exercise of capturing the unpredicted and using it to structure knowledge on the subject. Printed matter and reflection were at once sources of information and subjects for analysis and dialogic interaction with "informants" turned interlocutors.

Fieldwork was indeed scattered, or, to adopt Marcus's (1995) terminology, "multi-sited." This fact increased the proverbial field anxiety about methods, data, and the analytic objectification of a subjective experience. These minor worries overlapped with the more direct and brutal experience of witnessing the effects of the epidemic on people I got to know personally and became friends with throughout the period of research; this also sharpened my perception of the universality of the experiences brought by AIDS when it comes to death, fear, powerlessness, or willingness to act—all so similar to what I saw in New York. Some other experiences were different, however, and worthy of cultural analysis. These included differences in the style of social responses, especially in the interactions between society and government, between social organizations and the medical establishment, between patients and doctors, between people with AIDS and their treatments, and between the public and the information the media provided about AIDS. Those differences added to contrasting patterns of sexuality, drug use, blood distribution, and other transmission-related issues, which had already been analyzed by social scientists.

And yet those contrasts could not be fully analyzed through a cross-cultural comparison of responses to AIDS—for instance, New York and Rio. More than just being "cultural contrasts," the differences were mediated by the asymmetries of the world system. This problem, only implicit in my initial question, grew increasingly apparent as the analysis proceeded. In Brazil, AIDS was defined primarily as an international issue, and by that very fact the representation of the disease encapsulated the contemporary asymmetries within the world system. This aspect of the disease could not be left out of the analysis. Local commentary elaborated on the topic: Some Brazilian analysts noted, for example, that "AIDS-the-phenomenon" arrived in their country before "AIDS-the-disease" (see Chapter 4). Unlike local responses to other massive health problems in Brazil, the local responses to AIDS were influenced by responses in the United States and responses by international agencies. From the start, responses to AIDS in Brazil were relational—and not so much between the social movements and the medical establishment (as in the United States), but through the development of a permanent means of interaction with the international action and related networks.

That relational mode inspired the form and sequence of this text. After approaching the theoretical problem and related literature (Chapter 1), I give some background information on the international context (Chapters 2 and 3) and offer a separate treatment of the worlds of social responses (Chapter 4), medical settings (Chapter 5), and scientific research (Chapter 6) in Brazil. It became increasingly clear that interactions between any of these spheres and their international networks regarding the production and use of knowledge were more relevant than any interactions they might have with one another—even if they coexisted in the same city, shared personnel, and promoted joint projects. During the years of preliminary trips, fieldwork, and follow-up visits, I could observe the extent to which international references, relations, pieces of information, and direct interaction were part of local daily life. This happened in a fractured way, as each of the sectors had its own international network; it also had fracturing effects, by sending the different sectors into divergent directions for interaction. At a level of knowledge production and reproduction, the staff of ABIA, for instance, found that it was easier to make a connection with the Geneva-based staff of the Global Programme on AIDS than with the staff of local hospitals next door, a few miles uptown, or across Guanabara Bay. The social actors each had different, specialized knowledge, and they had only fortuitous contact with one another. Daily interaction across transnational spaces materialized in all sorts of ways: funding, networks, literature, references, connections, messages, mail, fax and electronic mail, and actual visits. Deeply embedded in the international response, local agencies were an integrating part of transnational networks, and fieldwork had to include their participation in international forums.

Fieldwork was certainly not confined to a small community. The subject constantly exploded beyond its limits, often onto a worldwide level. Unlike the physically and distinctively distant field sites of traditional anthropology, this setting was present everywhere and anywhere, a fact that increased the difficulty of finalizing fieldwork on a subject that was already unusually distressing. The text that follows is an attempt to disembody this experience using the formal language of social analysis, literature review, ethnographic accounts, and theoretical discussion.

Global Responses to AIDS

ONE

AIDS and Science

THE AIDS EPIDEMIC arrived at a time of exceptional optimism about the powers of medicine to solve human problems. There had been significant medical achievements in the previous decades: difficult biomedical challenges had been overcome, epidemics had been controlled, and infectious disease was no longer a technical problem. In the early 1980s, at least in the developed countries, it seemed as though we were to live longer and healthier than ever before. Furthermore, and under the mindset of modernization, this state of things might be extended to all of humankind. As infectious disease was now treatable, other illnesses might be so in the near future. Development could bring a world of health for all. "Health for All by the Year 2,000"[1] was actually the name of a series of publications issued by the World Health Organization (WHO), evidencing the optimism shared by this postwar international agency.

This optimistic perspective, coming from the inner core of biomedicine, was not matched by a sociological evaluation of the status of health around the world. There may have been an enormous number of medical achievements in the twentieth century, but their benefits were not available for all. High mortality rates persisted around the world, mainly among the poor and deprived. An inordinate number of people continued to die and suffer from technically treatable or preventable infectious diseases, in epidemic and endemic forms. Health patterns and life expectancies varied according to social and economic factors, including income, education, gender, race, age, and place of residence. "Health for all" was far from being achieved, since medical advances do not necessarily correspond to a widespread improvement in health. Those issues were more likely to be framed within the fields of public health and social sciences rather than at the core of biomedicine. The social dimensions of health were somehow marginal to the priorities of central medical research.

This symbolic hierarchy in the perceptions of the dimensions of health changed, at least in some instances, during the late 1980s and 1990s as a result of the effects of a new and unpredictable epidemic known as Acquired Immune Deficiency Syndrome, or AIDS. More than any other health crisis, this new epidemic required the combined efforts of those involved in technology-driven biomedical research as well as those in the social disciplines, bringing together the "inner core" of hard-data-driven science with the "fringes" of soft-data disciplines. AIDS raised a number of questions that had traditionally been part of medical anthropology theories or public health discussions but rarely, if ever, had been of central relevance to clinicians or biomedical researchers. Those questions became a priority for everyone trying to understand and manage the disease.

The perception of AIDS as an epidemic that would change the priority of many issues in medicine, as well as the relationship between medicine and society, did not come early, nor did it occur to everyone. Except for those who were directly affected by the disease, AIDS did not appear to be too important. During the 1970s and 1980s, the biomedical community felt that there was no room for a new major disease. Cancer was already believed to be the last frontier, consuming the leading energies and funds available for medical research. The thinking was that it, too, would be brought under control at some future moment. It was time for biomedical reports to celebrate the irreversible achievements of their field.[2] In one article, AIDS was depicted as a diversion from bigger tasks:

> It seems probable that new drugs and vaccines will at some future date bring [AIDS] under control. . . . in the interim, AIDS will consume the time and talents of many research scientists, and much money, and medical care resources will be *diverted* to care for those stricken with this "disease of social change." (Rogers 1986:221; emphasis added)

The authors seemed certain that full control of human health would become a reality. Cancer and the remaining "mysterious" diseases would supposedly be defeated sooner or later, just as infectious diseases had been.[3] The so-called new diseases might be topics of interest to the media, but they were not real medical concerns: they simply were not supposed to exist.

The "war on cancer" epitomized the choices, challenges, and direction of medicine in the 1970s.[4] The expression itself deserves analysis. "War" is a convenient metaphor for a massive attack against an identified "enemy," which may coincide with a social scourge. Politicians had already marketed the "war on poverty" in the 1960s and, later, the "war on drugs." "War" also suggests the mobilization of funds and energy. This metaphor is not used casually to describe approaches to problems in medicine and sanitation.[5] In the sense that it frames the direction of medical research, and affects decision making and clinical management, representation shapes the actual course of epidemics.

The defenses, attack mechanisms, and correlate metaphors of armies become real, reducing biology to a struggle between armed humans and invading microbes, the former in need of strength and a strategy to defeat the "enemies." In these terms, medical science is not just opposed to ignorance and disease; it is presented as in a state of permanent war against pestilence. Furthermore, this war is waged on several levels: humans against microbes, cells against viruses, immune system against microbes, drugs against bugs.

In accordance with that logic, medical scientists had been "victorious" in the "war" against illness on a number of occasions, the most celebrated of which was the battle against infectious diseases. Infections and the microbes that caused them had been "defeated" in the twentieth century by so-called medical "magic bullets," such as antibiotics,[6] or eradicated by vaccination, both consequences of developments in biomedical knowledge. The "conquest" of smallpox was a landmark that inaugurated a new era, in which the specialty of infectious diseases began its decline, as medical writer Laurie Garrett points out:

> Surgeon General William H. Steward in 1969 told the U.S. Congress that it was time to "close the book on infectious diseases," declare the war against pestilence won, and shift national resources to such chronic problems as cancer and heart disease. . . . Young scientists got the message that fields such as medical entomology, parasitology, and host pathology were not the paths to rewarding, productive careers. (Garrett 1992:825)

In contrast to the "conquest" of infectious diseases, "victory" over cancer seemed elusive. In spite of the large amount of literature produced during years of intense and exceptionally well-funded research, there was only limited understanding of the disease. These efforts resulted in improvements in prevention (for example, through early screening) and therapy, but no major breakthroughs leading to a *cure*—at least, not according to the "warfare" mindset.

In that context, the hypothesis that cancer might be caused by an infectious agent, virus or otherwise, was appealing.[7] As an infectious disease, cancer would be less mysterious; it might even be controllable. Attracting scientific interest and funding, the pursuit of cancer-causing viruses became one of the leading endeavors at the National Cancer Institute (NCI). The outcome of this research was modest, however. The only major finding of this "virus hunt"[8] was the first description of a human retrovirus, in 1980, by the NCI team headed by Robert Gallo. The virus was named human T-cell lymphotropic virus (HTLV).

HTLV was more important to basic science than to clinical medicine. Associated at first with leukemia and later with lymphoma, HTLV was not easily associated with any recurring disease. One exception was viral lymphadenopathy, which, according to one research group, occurred at a high rate in Japan.[9]

For basic science, however, HTLV was quite an interesting finding. Retroviruses had been known for only a few years, and were not considered to be potential infective agents in humans.[10] The virus's genetic composition—a single strand of RNA, which contains only half of the genetic information required for a cell to function—was considered too simple and too basic a biomolecule to sustain life.[11] The virus had generally been an odd concept for the life sciences, since it is neither alive nor completely inanimate.[12] "Regular" viruses at least contain DNA (two strands forming a double helix), and are able to parasitize living cells, "borrowing" from their host the cellular machinery necessary to separate the two strands of DNA to make RNA, which the cell will use to make more viruses—at its own expense. With only a single strand of genetic information, the retroviruses (also known as RNA viruses) were not expected to be capable of using the host's genetic material to replicate and cause infections. In fact, retroviral infection was not considered possible until the discovery of *reverse transcriptase,* the enzyme that allows single-stranded RNA to be incorporated into the double-stranded DNA of the host (i.e., "inverse" replication of RNA).[13] Once the viral RNA is part of the host's DNA, the genetic information originally contained in RNA can be *expressed;* that is, copies of the original retroviral RNA can be made, as well as copies of the protein coat necessary to encase the RNA to form a new virus. Years later, when the infectious agent for AIDS was identified as a retrovirus, scientists tried to create a "magic bullet" against AIDS by targeting the reverse transcription process. The first of those drugs would be AZT.

* * *

Such was the state of biomedical research when cases of Kaposi's sarcoma (KS) were first reported in New York City in 1981: the war on cancer was ongoing, retroviruses were new to biomedical research, and infectious disease had supposedly been banished from the United States. Kaposi's sarcoma was considered a rare cancer that, for some unknown reason, appeared in a pernicious and lethal version among gay men in New York. It would later be identified as one of the pathologies constituting the syndrome of AIDS.[14] Not surprisingly, doctors involved with the management of the first cases of KS that were seen in New York, such as Alvin Friedman Kien, M.D., suggested that a virus could be causing the new disease.[15] They became optimistic about the implications of this hypothesis in their research on cancer,[16] which was considered the last remaining challenge to modern medicine. The understanding of its etiology and mechanisms of pathogenesis would advance biomedicine and broaden the scope of knowledge about human biology.

If a new infectious agent was needed to explain the new disease, HTLV was a good candidate for the role. Most scientists had been pursuing a viral etiology for AIDS rather than the popular explanations of the day which blamed environmental factors, God's punishment, or lifestyle excesses. NCI researchers were especially interested in HTLV as a possible cause for AIDS. In their laboratories, HTLV had been a significant deprived of significance. AIDS might finally give the virus a meaning.

Gallo's hypothesis that HTLV caused AIDS was difficult to prove. In his team's laboratory, viral lineages perished fast, before any conclusions could be made.[17] His hypothesis was proved elsewhere—at the Pasteur Institute in Paris, where Françoise Barré-Sinousi was able to grow in culture a virus obtained from the blood of a patient with AIDS. In May 1983, she isolated the new virus and named it RUB, after the patient from whose blood it had been obtained. RUB was later renamed LAV for lymphadenopathy-associated virus.[18]

Evidence of reverse transcriptase activity in the "French" virus qualified it to be classified as a human retrovirus, like HTLV. However, its genetic structure did not match that described in 1980 for HTLV in the American lab. In 1983, French researchers sent a sample of LAV to the NCI to be compared with HTLV. Were they the same, which would prove Gallo's hypothesis that HTLV caused AIDS? Were they variants of a common family of viruses? Or were they completely different from each other?

Even though the French researchers considered LAV to be very different from HTLV and classified their finding as a *lentivirus,* the NCI researchers declared that LAV matched their own, newly discovered AIDS-causing HTLV III, which was supposedly a new variant of HTLV. It took years to find out that irregular laboratory procedures had led the Gallo team to mistake LAV as a variant of their own HTLV.[19]

At about the same time as his colleagues, California researcher Jay Levy isolated a strain of virus from the blood of people with AIDS and named it *AIDS-related virus* (ARV). Lack of funding prevented the Levy research team from participating in the "co-discovery" of the cause for AIDS.[20]

The discovery of the agent that caused AIDS—HTLV III—was announced to the American public at a media conference by Secretary of Health and Human Services Margaret Heckler, who referred to the Gallo team's discovery as "another miracle in the long honor roll of American medicine and science."[21] This announcement was made on April 23, 1983, just one day after the *New York Times* ran a story about the French team's discovery of an AIDS virus.[22] With their findings announced in the public media rather than in a scientific forum, the American scientists were made to appear heroic. The public announcement also raised the hope of finding a vaccine—and thus a cure for AIDS—within the next couple of years. Pinpointing a microbial enemy as the cause of the epidemic fit the "warfare" mindset and seemed to be the first step toward a cure. If nothing else, it allowed researchers to change the classification of AIDS from an obscure form of cancer to an infectious disease.

As an infectious disease, its epidemic pattern made more sense. A transmissible microbe would explain the clusters of cases that were found among urban American gay men, drug users, people from the Caribbean and parts of Africa, blood recipients, and infants. The specialty of infectious diseases was in the spotlight again, after years of virtual oblivion and peripheral developments—hospital patients, military medicine, or a survival of the old-fashioned tropical medicine that persisted in the Third World.

From then on, most of the initial research efforts were based on the assumption that AIDS was caused by an infectious agent; thus they were guided by the paradigm of germ theory: one microbe, one pathology, one treatment. Research on the new virus, soon to be renamed *human immunodeficiency virus* (HIV),[23] would soon join the molecular biology "bandwagon," to use Joan Fujimura's depiction of recent cancer research in the United States. Research was predominantly oriented toward knowing the "invading enemy" by bits and pieces, down to its genetic sequencing, so that an appropriate "counterweapon" could be developed.

Tension arose between a "molecularized" and a "multidimensional" approach to this disease. The fact that a state-of-the-art technical solution took a few years and much collective pain to be released gave rise to speculation and distress. Explanations varied from non-medical moral theories to explain the disease to the exploration of alternative approaches to care within the health and social sciences, including the social dimensions of illness, its psychological and emotional aspects, a closer understanding of the complexities of the infected immune system, and the variety of immune reactions, among many others.

The molecular approach finally yielded a pharmaceutical solution that pleased many and, at the time of this writing, has gained a medical consensus and social approval. This corresponded to a simultaneous "attack" on the different enzymes related to HIV's activity with specific antiretroviral agents, including the newly developed family of *protease inhibitors*. This strategy, known as *triple therapy* or the *antiretroviral cocktail*, was announced at the 11th International Conference on AIDS, held in Vancouver, Canada, in 1996. This announcement brought relief and hope for many. It had the effect of "normalizing" the AIDS epidemic, making it seem like a treatable—albeit complex—infectious disease. This also meant that AIDS could, like other epidemic and endemic diseases that persist in spite of the existence of a cure, join the other diseases of dual pattern: curable in the First World and lethal in the Third World.

* * *

The lapse in time between the identification of AIDS as a medical problem (the early 1980s) and the clinical/pharmaceutical response to it (in 1996) with combination antiretroviral drug therapy permitted a rare encounter between the medical and social perspectives. Since medical solutions to the epidemic were inadequate during that period, the social dimensions of the disease became more prominent, visible, and central. Social scientists, social workers, epidemiologists, and the people directly affected by the epidemic played major roles in framing the direction of biomedical research.

Presented as a medical and a social problem at once, AIDS bridged a gap that had traditionally been used to distinguish between strictly "medical" problems, such as cancer, and "social" problems like tuberculosis. When KS was identified as a possibly virus-induced cancer, it bridged that gap at a microscopic level, where the fields of cancer and infectious disease were actually closely related. When AIDS was found to affect both affluent and developing societies alike, it bridged a gap at a macro level: the gap between the two poles of a dual world health system.

That bipolar system derived from the interpretation of contrasts in morbidity and mortality statistics at a world level, which showed a dual pattern: a developed and "modern" one, "post-epidemiological transition," with high incidence of cancer; and an underdeveloped one, "pre-epidemiological transition," with the prominence of infectious diseases.[24] Cancer and heart disease corresponded to a burden of modernity, for they stood out where infectious disease had succumbed to adequate health care, nutrition, education, and economic well being—that is, in the developed countries. In the absence of social and economic well-being and other development indicators, the number of cancer cases in the more disenfranchised parts of the world was submerged among the higher number of cases of tuberculosis, malaria, cholera, schistosomiasis, leprosy, filariosis, tryponosomiasis, plague, and child dysentery—to name just a few of the technically treatable or preventable diseases that still kill and frighten populations in many parts of the world.

During the period when AIDS had no satisfactory medical response, these nontechnical issues were examined much more closely by medical researchers than they had been for any other disease in the past. The WHO AIDS program (see Chapter 3), for example, gave priority to the social dimensions of the disease in its approach to the epidemic by emphasizing prevention and education and addressing such social issues as politics, empowerment, gender relations, labor mobility, sexual commerce, drug trafficking, and the safety of blood supplies. Suddenly, problems that were familiar to health care workers in developing countries became concerns for those in affluent countries, who usually held a narrower and more technical view of medical problems.

Classified as an incurable infectious disease—or, rather, as a lethal combination of what used to be treatable infections—AIDS challenged the opposition between, on the one hand, the chronic and modern diseases of the affluent and, on the other hand, the infectious and lethal diseases of the poor. The pandemic brought to the developed world an experience of collective distress and helplessness that had long since been forgotten there. AIDS struck across divisions of class, nation, race, and gender, as documented by major health agencies,[25] and highlighted the contrasts and contradictions of contemporary society, as documented by social scientists and international agencies.[26]

This epidemic simultaneously resulted in *homogenization*—the bringing together of people from developed and underdeveloped countries—and *heterogenization*—that is, intensifying existing differences in class, gender, nation, race, and sexual orientation,[27] through differences in access to care and risk for disease. AIDS showed the contradictory face of the contemporary world: on one hand, an interconnected society, capable of distributing information simultaneously across the globe to achieve the illusion of proximity and shared experiences; on the other hand, a society in which global interconnectedness increases differentiation and contrast.

For that reason alone, the progression of the AIDS epidemic would be interesting to study on a global level as a means of understanding contemporary health problems and the interrelationship of the medical and social dimensions of illness. Furthermore, a study of the response to AIDS between 1981 and 1996 would be a unique case to study the social process that produces medical knowledge. Combining these two fields, our study introduces the variable *world asymmetries*, or the opposition *centers–peripheries*, into the current discussions on the social production of science.

* * *

One of the most clearly distinct features of the responses to AIDS was the fact that they were not produced solely in the isolated environment of a lab, in the ivory towers of academia, or during scientific conferences. From its earliest stages, AIDS research was subject to public scrutiny, lobbying efforts, discussion, and negotiation.[28] From the beginning, the public reacted with discontent and anger at the failure of an affluent society such as the United States, with its self-confident medical community, to respond with an effective and timely solution to the AIDS crisis. Critical writers and activists blamed the inadequate responses on lack of funding, political interest, basic science research efforts, social understanding—or simply on homophobia, since the majority of the visibly affected population was gay.[29] But even when more research, money, awareness, and government commitment were guaranteed, a prompt solution was not forthcoming. The absence of a unified and strong response on the part of science had as a counterpart the development of a multivocal panoply of responses—somehow consistent with the multilayered character of the crisis. AIDS was too many things at once: a new, multidimensional, devastating experience of grief and loss for individuals as well as for society; a focus for social activism; a scapegoat for conservative hatred; a basis for social intervention in developing countries and for international aid. For social scientists, the AIDS crisis was also the grand challenge of a "total social fact," whose whole we try to grasp while we keep producing bits and pieces of partial knowledge. But science, too, took its time to come up with a consensus, and the process that led to it left us room for scrutiny. A project of anthropology of science should examine this sequence, analyzing through the layers of knowledge and action produced in response to the crisis.

Responses to AIDS included a variety of "products," from literature to chemistry, from social movements to worldwide action. New scientific journals were created, such as *AIDS: the Journal of Acquired Immune Deficiency Syndromes,* and *AIDS Research and Human Retroviruses.* An extensive number of articles and special issues also appeared in pre-existing journals, such as the *New England Journal of Medicine,* the *Journal of the American Medical Association* (*JAMA*), *Annals of Internal Medicine, Lancet, British Medical Journal, Reviews on Infectious Diseases, Journal of Infectious Diseases, Science, Scientific American, Science et Vie,* and *Social Science and Medicine.*

Many books also appeared—not only textbooks of medicine, psychology, and social work, but also policy reports and recommendations, self-help books, and books on survival, life-enhancing strategies, activism, and social analysis. New popular magazines, either from the gay press or specifically for HIV-positive clientele also appeared, as was the case with *POZ.* There were also treatment newsletters, such as *AIDS Treatment News* in San Francisco and *Treatment News,* TAG's newsletter, and the handouts of ACT UP's Treatment and Data Committee in New York. In addition, a variety of other materials appeared, including videos, films, and books that addressed the social, political, and human aspects of AIDS in documentary or fiction formats. Emphasizing various aspects of the AIDS crisis, the mainstream media covered the epidemic by creating special interest columns and assigning science journalists to cover AIDS-related events.[30]

A new genre of literature gave voice to those who lived with the burden of this new and lethal disease that painfully exposed the limitations of contemporary medicine.[31] Expressions of grief extended beyond real-life literary chronicles to include the performing and fine arts, resulting in several projects dedicated specifically to issues concerning AIDS—*Gran Fury,* Keith Haring's works, Bill T. Jones' choreography, and the *Day Without Art* event—testimony to the devastating effects of the epidemic within the artistic community. Perhaps the most expressive testimony to the distress caused by the epidemic was the collective project *Names,* or The Quilt, which is composed of quilts, each of which was made in honor of someone who died of AIDS by his or her friends and family. The Quilt reached such an enormous size that it can only be seen in its entirety when it is spread over a large open area; for this reason, it is only shown on special occasions, such as during the AIDS conference in San Francisco (1990) or at The Mall in Washington, D.C. a few years later.

Development agencies produced special reports on the effect of the epidemic in developing countries. They were part of the international scenario, not just as recipients of foreign aid to cope with the epidemic but also by contributing to the expansion of our knowledge of AIDS.[32] Their experiences with AIDS, prevention strategies, and social agendas are stated in virtually every issue of newsletters such as *AIDS Action, Exchanges,* and the Brazilian publication, *Boletim ABIA.* International agencies presented their synthesis of the diversity of experiences with AIDS on their "global action" agenda.[33] New media were adopted to handle the rapidly increasing amount of information on AIDS, including CD-ROMs, faxes, on-line conferences, e-mail and other computer network devices (such as the electronic databases AIDSline and SIDSA, and a number of bulletin board systems on AIDS), and AIDS agency websites. Interactions among the media caused the number of circulating messages to multiply, making it harder to manage existing information and impossible to create a single structured account of AIDS.

Anthropologists and other social scientists contributed to the narratives on AIDS in various ways: with accounts from the midst of the epidemic;[34] with prevention-related or assistance-related research;[35] by compiling the challenges of producing theory and testimony in the midst of the crisis;[36] by studying responses to AIDS among specific social groups such as gay men,[37] drug users,[38] women;[39] or contextualizing the epidemic in particular socio-cultural settings such as Africa,[40] Haiti,[41] Brazil,[42] and the contrasting U.S. or European settings;[43] developing the social study of sexuality;[44] analyzing representations and cultural aspects associated with risk and stigmas;[45] questioning methodological and theoretical aspects;[46] or creating social commentary.[47]

Anthropological contributions did not change the fragmented nature of knowledge about AIDS. The involvement of anthropologists with the epidemic often located them within the local narratives produced by each experience of it, generating a diversity of perspectives that were not exempt from occasional antagonisms, such as those that arose among participants on the 1992 American Anthropological Association panel on AIDS,[48] or among the readers whose heated polemics appeared in *Social Science and Medicine* in response to the publication of a racist article and the authors of that article.[49]

Rather than exhibiting the old disciplinary trademarks of monographic synthesis and holistic overview of the topic under study, anthropologists studying AIDS were as dissonant and fragmented in their points of view as was the AIDS knowledge base at that moment. Their works portrayed the state of the discipline at that time. Would a deeper level of synthesis, in methods and in theory, be desirable from any perspective? During my research, I sensed that some of the AIDS-related social phenomena "demanded" a higher level of synthesis, one that would have to involve exploratory methodology and the development of theory. Such was the case for the multi-tracked International Conferences on AIDS (held annually until 1994 and biennially since then), WHO's multidepartmental AIDS program (*Global Programme on AIDS*), and the action of agencies such as the *Global AIDS Policies Coalition* (see Chapter 3). These phenomena intertwined science, clinical management, activism, policies, politics, issues of development, and social analysis. Their format seemed to suit the old holistic tradition of the discipline, with an added "macroanthropological" flavor.

Inspired by the pre-existing macroeconomy and macrosociology, a macroanthropology,[50] involving the study of global issues and taking its worldly field as an interconnected web of networks, could provide the framework to address the universe of AIDS in a synthetic manner. The interconnected social networks that handled the epidemic might be approached as a community that had some ritual physical appearance (such as the yearly mega-conferences) and occupied the entire world: a community of contradictions and dissonance, very much a post-modern presentation. A correlate approach might look like a collage of contradictory narratives portraying the history of AIDS as the ultimate post-modern topic: a scattered horizon of dissonant voices on the verge of producing either a pointless cacophony or an endless number of irreducible, self-sustaining truths—a chorus that was ultimately silenced by the stronger narrative of modern medicine's triple cocktail.

Rather than serve as the goal of this analysis, that "post-modern" scenario serves as a basis for inquiry into a "modern" issue—taking science as the unfinished project of the Enlightenment and of modernity.[51] Attributing to science the strength of a modern narrative, based on reason, not just any discourse among others, does not equate with naïve realism, much less with the conviction that science is outside the scope of social inquiry. Not only is the direction of scientific inquiry susceptible to the effects of all sorts of nonscientific (i.e., less "enlightened") factors—from the interests of private laboratories and academic or political institutions to the interests of competitors within the scientists' "star-system," to xenophobia, homophobia, racism, sexism, and often sloppy research[52]—but the very process of promoting a particular type of scientific inquiry is a social activity that deserves social analysis. In the study of AIDS, for example, the direction of scientific inquiry is the product of human activity and a complex mirror of social structures which are appropriate subjects for sociology and anthropology of knowledge.[53]

Defined as attempts to understand the relationship between specialized bodies of knowledge and the cultural and social contexts in which they are produced,[54] the fields of anthropology and sociology of knowledge have traditionally left science out of their scope.[55] One of the reasons for this exclusion is science's self-definition as universal knowledge and its claim to stand above the constraints of local culture. The traditional argument regarding the difficulty of studying the production of scientific knowledge from a social science perspective is expressed in terms of the particular epistemological status of science,[56] which is the same status that grants legitimacy to sociology and anthropology. Or, in Bourdieu's words, the difficulty in developing a sociology of science (his use of the term *sociology* is extended to include the social sciences in general) lies in the fact that "the sociologist has a stake in the game he undertakes to describe."[57] In fact, it is a double stake—one of the scientific nature of sociology in general, and the other of the particular form of sociology that is employed.[58]

In recent years, however, there has been a growth in interest in theory within the field of Science Studies (also known as Social Studies of Science [SSS], the Sociology of Scientific Knowledge [SSK], Science and Technology Studies [STS or S&TS], or the Cultural Study of Science [CSS]).[59] Scholars working in these areas have documented for several disciplines what medical historians and medical anthropologists have been saying all along about medical knowledge: that it is socially produced and historically situated.[60] Philosophical scientists[61] previously approached the issue of the social construction of scientific knowledge in their own terms,[62] thereby relativizing science's claims of absolute universalism. Scientists, too, have used the tools of social and historical analysis.[63] Ludwick Fleck's work on syphilis and the Wasserman reaction, published in 1935, is a paradigm on which many of the fields involved in the social studies of science—particularly the biosciences—are based.[64]

The most well-known version of the social study of science is the Mertonian sociology of science.[65] Robert Merton focused on the external sociological aspects of science and suspended the discussion on the actual contents. In Latour's words, "recognition, quotations, competition, budgets, were what impassioned American sociologists [studying science]."[66] The study of the cognitive "internal" result of such activity—that is, scientific knowledge—was left to humanistic disciplines such as philosophy, epistemology, and the history of scientific ideas.

Merton is considered a disciple of Talcott Parsons's Weberianism; however, epistemological externalism is not a necessary outcome of Weber's theory. In his Sociology of Religion,[67] there were enough guidelines to carry out a creative social exploration of the contents of knowledge, including scientific knowledge. Guidelines are also provided in the classic works of Durkheim,[68] Marx,[69] and Mannheim,[70] and in the works of anthropologists[71] and constructivist sociologists,[72] which before the 1970s addressed the cultural and social dimensions of science.

In spite of these preliminary efforts to account for science within the scope of the sociology of knowledge, it wasn't until the 1970s that there was a full attempt at a sociological exploration of scientific knowledge. Pierre Bourdieu's 1975 analysis of the scientific field helped to break ground in this arena by eliminating the division between externalists and internalists. In his own words:

> An authentic science of science cannot be constituted unless it radically challenges the abstract opposition . . . between immanent or internal analysis, regarded as the province of the epistemologist, which recreates the logic by which science creates its specific problems, and external analysis, which relates those problems to the social conditions of their appearance. (1975:22)

A few years earlier, British sociologist Michael Mulkay (1972) approached the question by discussing the social conditions for paradigm shifts or scientific revolutions, which the epistemologist Thomas Kuhn had described as being products of the internal proprieties of the scientific process or of the pure interplay of ideas.[73] Placing the analysis within the "problem networks" that constitute scientific communities, Mulkay rightly asked that the next step of inquiry focus on the effect of social factors outside scientific networks in shaping the contents of science. Also, the presence of Foucault's *Archaeology of Knowledge* (1969) in 1970s intellectual life provided know-how and a background reference for denser explorations of the social and political character of knowledge constructs.

Whether they were influenced by these works or by a broader historical moment and a shared *zeitgeist*, a few sociologists and anthropologists broke what had formerly been a "taboo" by examining the production of scientific knowledge directly during the late 1970s and early 1980s.[74] The endeavor was pursued on different fronts. The so-called new sociology of knowledge combined ethnographies of the laboratory, discourse analysis, and detailed accounts of the process of achieving consensus in science to produce a thick description of the micro-social processes through which the contents of science are produced and the identities of scientists are defined.[75]

Another important source for the social study of science is the feminist literature, particularly literature that contains a critical examination of hegemonic discourses and practices. Feminist authors challenge male authority in scientific models in a number of ways.[76] This approach became a central reference and inspiration for the social studies of science in the United States.[77] Even though its transformative role is rarely acknowledged by philosophers of science,[78] this stream of thought is the favorite target for "realist/rationalist" opponents of the sociological analysis of science.[79]

Based on a different yet closely related social motivation, analyses that question the scientific basis for racism also challenge this form of scientific discourse and the public policies derived from it, as Reid points out when referring to the case of sickle cell disease in the United States.[80] Far from being buried as a historical reference, scientific racism makes frequent comebacks, as demonstrated by the recently published book *The Bell Curve*.

The social study of science is complemented by other methods of exploring scientific practices and productions that come from diverse sociocultural contexts.[81] Recently, the multicultural dimension was added to the discussion on the identity of science.[82]

Today, social studies of science gravitate to aspects of technoscience and technoculture,[83] and include topics from cyberculture[84] and artificial intelligence[85] to artificial life and reproduction[86]—as well as contemporary medical topics such as genetics,[87] the formulation of questions,[88] the immune system,[89] AIDS,[90] medical technology,[91] the limits of humankind and cyborg anthropology.

Even though most of the authors of recent works that examine scientific knowledge came to terms with relativism, some are still uncomfortable with it—mostly outside observers, particularly those who equate science with unquestionable realism. Most recently, that discomfort exploded, resulting in the so-called *science wars*.[92]

But the tension was there before. Not necessarily a consequence of the analysis, relativism lingers around it, as researchers in social studies of science have debated. By opening the "black boxes" of science[93] and exposing, through analysis, the "guts" of the social process involved in constructing scientific knowledge, social scientists have developed a meta-knowledge that may not be welcomed by all. Latour recommended that this information should not be made accessible to the practitioners of science, since they perceive their activity as being incompatible with the relativism that stems from the analyses. For anthropologists, the problems of relativism are all too familiar. Too often we are placed in borderline situations where local ethics and beliefs clash with our own. For medical anthropologists, who must address issues of life and death, relativism takes stronger forms, as is seen in the study of AIDS.

Life might be easier under unquestioned realism; in the study of AIDS, however, there is too much evidence against the unquestioned status of scientific products, which appear like a shifting ground where statements replace one another at a very fast pace, with each statement turning out to be as fragile as the one it replaced. In this chain of fragilities stand public policy, medical decisions, and individual choices—all with a close and direct effect on matters of death or survival, treatment choices, quality of life, sexuality, fear, prejudice, support, prevention, political intervention, and international action. The fact that absolute decisions depend on a disproportionate number of relative and temporary certainties only increases the distresses of relativism.

The fact that this particular body of knowledge has been so closely scrutinized means that a large body of documentation and data exists that exposes the fragility and transient character of the products of scientific research. Public scrutiny also increases the number of factors reverberating in the final consensual results; public discussion is not just the poorly tolerated "ghost in the machine"[94] whose existence is acknowledged but not incorporated into other kinds of knowledge. Bioscientists involved in AIDS research cannot ignore public discussions and social scrutiny of their work; these social forces directly influence research by putting pressure on research funding organizations.

There are other problems in the study of society's responses to the AIDS epidemic. We face a body of information that is obtained from multiple and contradictory social contexts and controlled by social mechanisms that rapidly scrutinize the process by which it is obtained and provide feedback to the communities from which it was obtained. Our knowledge of AIDS reflects this multilayered, multipartite process and gives us clues about contemporary society while reminding us of the social character of the production of information in the biosciences. In this study, what is known about AIDS is presented as a composite of social knowledge, activists' knowledge, public awareness, academic production, and clinical experience. The process involves a range of social actors, from scientists and physicians to activists, public officers, non-governmental organizations, and international agencies. I will describe the actors and the social process of knowledge making with a particular focus on its social negotiation,[95] and the final outcome in fields such as prevention, treatment, epidemiology, and immunology.

Last but not least, there is a further challenge for the theoretical ambition of this book: to introduce a *world-asymmetries variable* into discussions about the social construction of science, as illustrated and highlighted by the facts of the AIDS epidemic. The AIDS crisis helped make visible a number of different social issues.

* * *

Among other things, the fact that this disease brought the attention of the main centers of research back to infectious diseases (which were previously declared to be under control) has changed the terms of center–peripheries interactions. The peripheries (usually the Third World) were better acquainted with infectious disease and with the social dimensions of illness than the centers (usually the First World). Is it now time for the centers to learn from the peripheries instead of solely exporting information gained from their experiences, disregarding the possibility that their social actors can be creative producers of knowledge?

The very idea that developed centers could learn anything other than raw data from the underdeveloped margins was nearly inconceivable before the advent of AIDS. The understanding of science as part of modernity is so consensual that the modernization paradigm is the reference used to explain the unequal distribution of scientific production in the world. There would be no place for scientific production in the peripheral countries. Science goes with modernity, and that is a propriety of the few "developed," central countries. Arguments designed to reinforce this belief include the following:

- Scientific development requires the resources and infrastructure that only developed countries can provide.
- The few existing scientists in developing countries are either frustrated by the inadequate working conditions in their home coun-

> tries and do not produce enough original research there or migrate to research centers in the developed world, thereby becoming a part of a cosmopolitan society rather than the developing world. Without local production or funding, developing countries could not contribute to new scientific knowledge.

This ideology reinforces the invisibility of possible scientific production in peripheral settings, a theme that has been raised by scholars from these countries, particularly from Asia[96] and Latin America,[97] as well as more or less radical analysts from the United States and Europe[98] and others who conduct research sponsored by the United Nations.[99] They write about "underdeveloped science," "science and modernization," "dependent science," or "Third World science." Often their perspective is inspired by dependency theory; they argue that the scarcity of scientific production in peripheral countries is not a sign of backwardness but an effect of the same mechanisms that cause underdevelopment. In other words, the problem is not that scientific creativity is absent in the developing world, but that such creativity is suffocated either by research interests that are defined elsewhere, by the economic interests of the local elites, by the Eurocentric epistemology of mainstream science, or by a complex interaction of socioeconomic and political factors that reinforce dependency.

It is no coincidence that a significant portion of critical literature about dependency comes from or is inspired by the experience of Latin America after World War II, just as the most current discussion on post-colonialism is inspired by literature created in recently decolonized South Asia.

In Latin America, the externally induced obstacles to internal development can be traced clearly. Another factor that makes Latin America (particularly the large and complex countries such as Brazil or Mexico) a very interesting place to examine center–periphery interactions in science, especially in the health sciences, is the proximity of the minority of literate and cosmopolitan middle and upper classes—who share social patterns and lifestyles with the developed world—and the majority of the population, who are poor and endure the burdens of underdevelopment. This social structure helped create a field for developing theories to account for the social and structural dimensions of nearly everything, including human health, illness, and medicine. Within the field of health sciences, for instance, some original theories evolved, such as the one espoused by the radical Latin American school of social epidemiology.[100] This type of epidemiology has been updated and made more syncretic in Brazil,[101] taking into account the flexibility of central locations and the constant production of peripheries.[102]

There is also the fact that infectious disease in Latin America is far from being "defeated." Rather, it is a major field of medical intervention, and precisely one in which the social dimensions of disease could not be more clearly demonstrated.[103] Dependency theory influenced many other aspects of the study of health in Latin America. In a country such as Brazil, it is fair to say that the entire medical field is influenced by the notion that, on some level, illness is socially determined. The sophistication of local knowledge in the field of social medicine and infectious disease contributed to my selection of that country as a counterpoint to the United States for the study of responses to AIDS and the ensuing knowledge production.

In addition to this reason, which will be explored further in later chapters, Brazil deserves special attention as a country that is severely affected by the AIDS epidemic. The contrasts and similarities with the epidemiology of the United States and the developed world, and other contrasts and similarities with African epidemiology, along with Brazil's internal contrasts and intertwined politics, make the case of AIDS in this country worthy of detailed study.

Under those circumstances, it might be that if there were cases of possible traffic and interactive negotiation of knowledge between centers and peripheries, Brazil was likely to be one of them. It deserved *in loco* analysis, and there were some clues about local creativity and center–periphery dual interactions in some areas of knowledge and in public responses to AIDS (see Chapters 3 and 4). Brazil might illustrate the fermenting power of encounters between different streams of knowledge, an effect described for disciplinary migrations and encounters between unusual partners.[104] Such encounters have been recommended as a strategy to overcome the limitations of the current knowledge of AIDS.[105] The encounters, forced by AIDS, between high-tech and peripheral medicine might have that "fermenting power." In a way, Brazil was already a site where internally those encounters had resulted in fertile social medical thought. Could anything equivalent to such encounters be developed on a wider, perhaps global, field?

An exceptional example of the "fermenting power" of encounters between various types of communities was provided in the United States, where partnerships developed between scientists and activists, doctors and patients, the gay-inspired social movement and the medical establishment, and pharmaceutical laboratories and community organizations to forge more efficient and comprehensive medical responses toward AIDS. These partnerships were hardly conceivable before this epidemic (see Chapter 2).

Another unusual partnership, sponsored by international agencies, brought together the First and Third Worlds, developed and developing nations, central and peripheral settings. Institutionally sponsored global action helped raise expectations of major change (see Chapter 3). Officers from WHO's Global Programme for AIDS (GPA) even mentioned the "health revolution" created by the responses to AIDS, which expanded the scope of action beyond the strict sphere of biomedicine to encompass its social, economic, and political dimensions. The notion of "revolution" helped mobilize the energies of social movements around the world and introduced to the field of AIDS a political activism that had developed around other social issues, such as the fight against military dictatorships in Latin America, community organizing, and promoting outreach to low-income populations.

This "globalizing" move resulted in the creation of an arena that united the ultra-developed and the underdeveloped, thereby overcoming the partitions between nations and bringing together the high-tech frontier of research and the "wastelands" of infectious diseases. What should be the impact of this junction in the ultimate universalist project of science? Was this "global" scenario the setting for the "truly universal culture" that might replace Eurocentrism?[106] This is not just an abstract theoretical point: given the interest in epidemics and infectious diseases in developing countries, globalization might be the key to revolutionizing current ways of thinking about AIDS.

Revolutionize. For a long time, it seemed that a leap in knowledge that was worthy of being called revolutionary was needed to overcome AIDS. The more that biomedical research appeared to lead to frustrating dead-ends, the more it seemed that a real change in health sciences was needed, perhaps along the lines of a Kuhnian "paradigm shift"—that is, a transformation in its *normative* rather than in its *descriptive* reading.[107] If encounters between partners with different backgrounds could lead to a shift in knowledge-base models, then AIDS seemed to be the occasion for such a change. The potential for these new partnerships to transform medical knowledge will be discussed in upcoming chapters of this book.

Outside the sphere of biological sciences, a significant shift in knowledge took place. It corresponded to the inclusion of social variables in the study of the disease and the understanding that its "eradication" would require profound social changes in macro and micro power relations (see Chapter 3). Further evidence of change is apparent in how biomedical knowledge is organized, with the growing trend of using interdisciplinary settings and transdisciplinary efforts.[108]

The "softer" discipline of epidemiology was an arena for change, but was marked by some inconsistencies in logic (see Chapter 3). In the early stages of the epidemic, epidemiology was based on a number of variables that were relevant to developed countries, which stressed the risk of AIDS in specific social groups that were defined in terms of individual characteristics, such as homosexuality and drug use. In the developing world, where the occurrence and outcome of illness was explained by social variables such as poverty, deprivation, labor migration, and explosive urbanization, the study of AIDS might have been based on a different set of variables.[109] However, evidence from the developing world was not used to make AIDS-related epidemiological theory or methodology. Variations in the epidemic patterns in African countries, for instance, were "domesticated" through the construction of an idiosyncratic, heterosexually defined "African" pattern II (see Chapter 3).

Any debates about the development of models used in the study of human biology and medicine have yet to be analyzed. Would these disciplines go beyond a narrow path defined by germ theory and molecular genetics, which seem to dominate the study of human biology and, consequently, the study of AIDS? Which "more universal" social construction could replace those models? And which social actors were to be involved in that change?

These questions will be pursued by approaching the interactions between centers and peripheries in the field of science. After describing how encounters between scientists and community organizations were the locus of innovation (Chapter 2), I will explore the possibilities of another potential, transnational partnership: the one between the front runners in medical research, such as elite laboratories engaged in microbiology and cancer research, and experts on local knowledge who have been involved all along with infectious disease (e.g., specialists in tropical medicine who work in developing countries). This center–periphery encounter—rather than more experimented analytical categories such as gender, class, or race—will be our key social variable in the study of AIDS.

Addressing this question from a social science perspective assumes a continuous "dialogue" between two sets of literature: literature that covers the open-ended study of dependency, world systems, globalization, and the aftermath of colonialism, and literature that covers current discussions on the social construction of scientific knowledge. This perspective will be taken from a particular peripheral region within the world system—contemporary Brazil. Brazilian bioscientists and practitioners from other fields—from tropical medicine and epidemiology to the social sciences and AIDS activism—rather than just a subject/object of anthropological analysis will be the partners and interlocutors of this approach, as well as its sources and points of reference.

TWO

Politics and the Construction of Knowledge

AIDS Activism as a New Social Movement

MOST OF THE SOCIAL COMMENTARY about AIDS has been permeated by the idea of its uniqueness.[1] Stigma, despair, drama, attention, morbid glamour, fear, action, politics, global panic, and global responses, on top of a peculiar biology, complex pathology, and odd epidemiology accounted to make this disease so special. But particular emphasis should be given to AIDS' engaged politics, which are at the root of its visibility.

This chapter is dedicated to discussing aspects of AIDS politics. Before the global dimensions of the epidemic can be examined (see Chapter 3), and before the intertwining relationship of medical, social, and political variables associated with AIDS in Brazil can be explored (see Chapters 4 to 6), we should discuss the first and most vigorous social movement generated in response to AIDS. This movement grew within urban gay communities in the United States in the 1980s and had an impact on several AIDS fronts: basic biomedical research, the pharmaceutical industry, local politics, and global policy-making decisions.

A few years later, communities in several other parts of the world faced the epidemic. Because every setting is historically unique, with an idiosyncratic combination of particular health problems and social configurations, each local response to the pandemic was an original combination of international elements and local characteristics. Gay-based AIDS activism in the United States became one of those international elements. Directly or indirectly, it influenced the responses to AIDS throughout the world—whether or not local gay communities existed. For that reason, before we can analyze globalization efforts or AIDS in Brazil, we should examine in more detail AIDS activism in the United States.

AIDS was not the first epidemic in history to trigger a social response.[2] From leprosy[3] and bubonic plague[4] in the Middle Ages to cholera,[5] syphilis,[6] Spanish influenza,[7] yellow fever, and smallpox[8] outbreaks as late as the early twentieth century, epidemic diseases have had significant social and political consequences in more than one time and location; the history of infectious disease is imbricated with the history of humankind.[9] Nor was AIDS the first disease to generate the formation of patient groups; people with diabetes, high blood pressure, cancer, and a number of rare diseases have sporadically formed their own support or advocacy associations.[10] During the 1970s, in reaction to a patronizing and patriarchal bias in established medicine, some feminist groups formulated alternative views and responses designed to enhance women's knowledge of their health and their bodies.[11] This strategy of empowerment included the development of networks, channels of communication, workshops, literature, and a commitment to make information available. Indirectly, this movement paved the way for developing the strategies used by AIDS activists.[12]

But the level of politicization of the patient-based activism that grew in the fight against AIDS was unprecedented. The social context of the epidemic was also unique. Its early identification as a "gay disease" defined the setting for the emergence of a new social movement that would influence responses to AIDS throughout the world.

AIDS was associated with gays from the very first time the epidemic was noticed. The first two articles on this new disease were published in the *Morbidity and Mortality Weekly Report* (MMWR),[13] a bulletin published by the U.S. Centers for Disease Control (CDC) in Atlanta, Georgia, to provide epidemiological updates and general surveillance information on disease trends around the world. This bulletin was a faster and more efficient way to spread the news than a traditional peer-reviewed medical journal.[14] Behind those articles and the early awareness of the disease was the preoccupation of the physician authors Michael Gottlieb from Los Angeles, Paul Volberding from San Francisco, and Alvin Friedman-Kien from New York with the occurrence of rare diseases in patients who happened to be young, otherwise healthy, and gay. Those young men were dying from the previously rare *Pneumocystis carinii* pneumonia in the wards of the University of California at Los Angeles Medical Center and consulting dermatologist Friedman-Kien at the New York University Medical Center because of odd lesions characteristic of a previously rare cancer called Kaposi's sarcoma (KS).[15] Later in the same year, several other clinicians mentioned the prevalence of unusual health problems among young male homosexuals.[16]

The word "homosexual" was dropped from the title of the first MMWR article on the new disease[17] but remained the most distinctive feature of the disease. Patients with mysterious cases of lymphadenopathy, immunosuppression, and strangely lethal versions of common infections—all forming a pattern that might be characterized as a *syndrome*—were not transplant recipients or leukemia patients; nor did they have genetic disorders or suffer from severe or chronic malnutrition, and nor were they elderly. The only characteristics they seemed to have in common were that they were young, male, and homosexual. This presentation remained associated with the new disease and influences almost every aspect of AIDS to this day.[18]

The AIDS–homosexuality link had two sorts of implications. The first was related to prejudice and to the synergistic interaction between homophobia and fears of disease and death: Deeply embedded horrors of contamination and contagion added to the general prejudice surrounding lethal diseases and were magnified in turn by fears and prejudice associated with differences related to sexual lifestyles.[19] Patients with "unusual symptoms" were often rejected and neglected by health service providers and in workplaces and residences, as well as by family members.[20] People were secretive about their health and serological status, and individual and collective denial limited the practice of preventive measures. In Brazil, the term *de risco* (of risk; at risk) became loaded with stigma and guilt and quickly turned into a derogatory expression.[21]

Even though epidemiologists established other categories of risk for AIDS, homosexuality was most closely associated with the disease. The stigma induced by homosexuality was passed along to patients in newly defined high-risk categories. Injecting drug users became equated with inner-city pariahs, as deserving of their illness as "sexual perverts" were. The manifestation of AIDS in the poorest communities in Haiti managed to stigmatize an entire nation. And hemophiliacs were already sick by definition. As the dynamics of the epidemic and its conceptualization caused some categories to be added and dropped, Haitians ceased to be treated as a risk group, women and infants were added, and more blood recipients became visible. At the same time, the general public and the media indulged in making a distinction between the "innocent" victims of AIDS (women, children, and transfusion recipients) and the "guilty" ones (drug users and active homosexuals),[22] with their distinction based on moral judgments similar to those that have been made about the "deserving" and the "undeserving" poor.

The social implications of each of these definitions of "risk group" would deserve lengthy commentary, but it was the association between AIDS and homosexuality that shaped most of the responses to the epidemic. The same link that enhanced stigmatization and dismissal and delayed research funding actually had a positive effect. It helped at least one community seriously affected by AIDS to channel its energy into the fight against the disease. The gay community used the same type of energy and mobilization efforts to fight AIDS that it had used in the previous decades to build the gay rights movement. It used its organizing know-how and community support to start a collective movement that changed the face of the epidemic and expanded the possibilities for anyone who was directly affected by it.[23]

The association of AIDS with sexual transmission and with a specific type of sexuality eventually motivated institutions throughout the world to support research on sexuality. It also introduced to the mainstream medical community and those responsible for formulating public policy what had once been the esoteric language of "gay-ghetto" culture. Furthermore, the AIDS–gay link influenced the fight for civil rights based on sexual orientation in settings where these were hardly conceivable. In many countries, the international fight against AIDS helped strengthen small local gay groups that could implement prevention campaigns and build community support for the sick.

Social scientists have shown that the category "gay" may be inadequate and ethnocentric.[24] In spite of the fact that homoeroticism and same-sex spontaneous or ritual practices have been documented for virtually every society and every historical period, a distinct identity for homosexuals is available in only a few regions of the world, and often for the purpose of discrimination. The use of the category gay/homosexual for purely descriptive reasons within a medical context (as in the initial MMWR reports) could hardly be found outside a few American sectors after the 1970s—that is, after the high point of the gay rights movement. The social category "gay" as we know it today resulted from the history of gays' struggle for civil rights and became a component of American culture beginning in the 1970s.[25] This was not the case in other countries, where "gay" is not spontaneously used as a descriptive category in medical literature. Randy Shilts commented on the fact that French researchers who dealt with the first cases of AIDS in their country considered it an "African disease" and considered the label "homosexual" for the epidemic to be very American.[26]

In fact, it was in the United States that the category "gay" became a symbol for individual and collective pride. This happened after more than a decade of struggle by gays and lesbians for their civil rights—a struggle that was modeled on the struggle for civil rights by African Americans and inspired by the women's movement. Although the process did not erase homophobia—just as the black movement did not erase racism and the women's movement did not erase sexism—the gay movement at least created the possibility of removing the element of prejudice from the term and associated it with positive feelings within specific contexts. There were now environments where "gay" was the norm and in the majority—if not exclusively by denomination.[27] In medicine, homosexuality was no longer categorized as a mental illness, and diseases affecting homosexual men were described candidly and clinically in medical textbooks.[28]

* * *

By the end of the 1970s, gay activists had been vocal for more than a decade in communities where the epidemic was most visible—San Francisco and New York City. There they had created social visibility; claimed individual rights and dignity for their lifestyle; fought discrimination, prejudice, and homophobia; and attracted gays from all over the world. In one sense, the effort to make one's homosexual identity visible and viable went hand in hand with a naturalization of the sexual/social category. Explicitly or not, and mostly to oppose the blame theories condemning homosexuals for their acts, many within the gay movement assumed that sexual orientation is an innate characteristic rather than a choice that can be made at will.[29]

This belief supported and applauded the "gay gene" research,[30] carried out supposedly to provide empirical evidence of a biological basis for homosexuality. It may seem paradoxical that biological determinism is welcomed by the same community that worships Foucault[31] and that once sought full emancipation from the chains of determinism.[32] These contradictions only reveal the internal diversity of the gay movement, within which libertarian anarchists coexist with conservative Republicans, and all sorts of ideologies and theories are available.

That such political diversity is held together by claims of a common identity defined by individual characteristics seems to be consistent with concepts held by social and political movements of past decades in urban centers of the United States. The women's movement and the black movement were also based on a collective identity, within which existed class stratification and differences in ideology, political orientation, and lifestyles. Internal diversity did not weaken those social movements, which were externally defined as "lacking" something and which reinforced group identity through either/or oppositions (i.e., black/white, women/men) that were originally at the root of prejudice and discrimination. A similar opposition—homosexual/heterosexual or gay/straight—became central to identity in the gay movement. Seen from the outside—for instance, from Brazil, where a number of nuances in sexual lifestyle between heterosexuality and homosexuality are available (as are "intermediate tones" between "black" and "white")—an absolutely dualistic system seems strange.

It is also odd that this dualism prevailed in spite of the decades-long availability of the Kinsey reports,[33] which portrayed human sexuality and sexual orientation as characteristics on a continuum, on which exclusive homosexuality and heterosexuality were just the extremes, and which we traverse with a certain amount of fluidity. As politics structured ideology and identity, independent gay communities became a social reality as much as a historical product, absolute in and of itself, yet simultaneously contingent on history. Both facts must be considered as we examine the explosion of AIDS and its social, political, and epidemiological consequences. The early identification of AIDS in gay men can be seen simultaneously as an epidemiological fact and as a social construction. This double characteristic became imprinted in the way AIDS was addressed throughout the world, whether or not it was related locally to male homosexuality (see Chapters 3 and 4).

The independence of urban American gay communities during the early 1980s could hardly be found anywhere else, except in a few Western European countries. This state of affairs would change with the development of responses to the epidemic and efforts to implement prevention strategies worldwide.[34] At that time, however, the spectrum of sexual categories and sexual behaviors, choices, practices, and ideologies was, for most of the world, arranged differently than in the United States, where the bipolarization between heterosexuality and homosexuality prevailed.

Recognizing both the lack of data on human sexuality[35] and the need to understand it better in order to promote AIDS prevention,[36] academics pleaded for funding and conducted a number of studies on sexuality. These studies illustrated the diversity of sexual categories, behaviors, ideologies, and practices across various cultures around the world, as well as their implications for the development of AIDS prevention programs.[37] Initially reifying the pre-defined high-risk groups as if they were independent cultural systems, these studies showed variation across and within established categories of analysis, whether they referred to ethnic minorities or to transmission groups, such as Latinos, African Americans, women, drug users, gay men, or blood recipients.

One of the concerns of individuals carrying out this line of research was to account for bisexuals and other men involved in high-risk behaviors for HIV infection who did not identify with any of the traditional high-risk groups or target populations.[38] Research findings showed that the category "gay" does not account for all same-sex behaviors, and that by targeting gay men exclusively through prevention programs, they would not reach many other men who were, in fact, at risk. Anthropologist Richard Parker and others argued that in Brazil a rhetoric addressed to gay men, which lacked culturally specific adaptations, would not be recognized by a large number of men who structure their (homo)sexuality under other lines of identity, such as activity/passivity or transgression.[39] Some Brazilian activist organizations, such as Associação Brasileira Interdisciplinar de AIDS (the Brazilian Interdisciplinary AIDS Association, also known as ABIA) (see Chapter 4), asked for a more radical recategorization of sexual groups and prevention targets. To support this effort, a number of Brazilian social scientists developed AIDS-related sexual research.[40]

The acknowledgment of such variation helped replace the static terms of identity with more dynamic ones based on behavior. Evidence of that shift can be seen in a chronological reading of World Health Organization (WHO) documents on AIDS. In 1988, the category "homosexuals" appeared in WHO documents. One year later, WHO acknowledged the limitations of that category and replaced it with the more descriptive "men who have sex with men."[41] A similar shift, designed to avoid the prejudicial nature of certain terms, involved replacing the term "drug abusers" with "injecting drug users."[42]

AIDS-related studies on sexuality, though numerous and extensive, did not constitute a unified corpus with methodological consistency. In the early stages of this research (the second half of the 1980s), they failed to account for the role of AIDS intervention programs in the legitimization of gay identity in settings where that identity did not fully exist. A second wave of studies included the important variable of AIDS-related gay activism and tried to explain its role in the full sexual landscape of AIDS prevention programs.[43]

* * *

Even though the social response to AIDS capitalized primarily on gay activism, the prevalence of the disease in the gay community was initially denied,[44] as the sector of this group that was affected first was not necessarily a political one. The first to be affected by AIDS were more likely to have been in the glamour circuit, in bathhouses, in clubs. Politics were not attractive for them.[45]

The initial years of the AIDS epidemic were characterized by a gap between the actual start of the epidemic and awareness of its existence. A variety of coexistent theories and explanations of the disease confused people. Many interjected self-blame when blame-the-victim attitudes were projected by the so-called Christian Right. Conspiracy and genocide theories coexisted with the theories of the radical right. Rumors of biological warfare, contaminated swine flu injections, chemically induced illnesses, and other popular fantasies also prevailed. Some authors became the voices of dissent as scientists showed that they were having difficulty arriving at a consensus on how to approach the disease.[46] Some of these voices made themselves heard through the popular press. *New York Native* editor Charles Ortleb became notorious for consistently dismissing not only the HIV–AIDS link but the entire concept of AIDS as a separate disease or syndrome, leading the paper into a systematic opposition to the AIDS mainstream. Years later, in 1990, *Native* writer John Lauritsen dedicated his book, which described AZT as "prescription poison," to a page-length list of "AIDS dissidents," those who challenged the HIV–AIDS link and the mainstream AIDS establishment.[47] In 1994, *Spheric*—the student newspaper of the City University of New York and the State University of New York—carried a special cover article entitled "Is AIDS a Hoax?"[48] The article questioned the HIV hypothesis, quoting the anti-HIV theories of retrovirologist Peter Duesberg, who was known for his opposition to the currently accepted theories.[49] The group HEAL, working out of the Gay Community Center in New York, also adopted anti-HIV theories and supported Duesberg's views. Almost fifteen years into the epidemic, they welcomed the fervent anti-HIV crusader Robert Willner, M.D., who was known to target the AIDS establishment in general and the NIH's Robert Gallo (see Chapter 1) in particular.[50] Claiming that AIDS was just the lumping under a new name of many diseases associated with immunosuppression (an idea shared by other dissenting yet more scholarly authors),[51] Willner used the flashy title *Deadly Deceptions: The Proof That Sex and HIV Absolutely Do Not Cause AIDS* for his book. He discouraged people with AIDS from taking AZT and accused the perpetrators of viral hypotheses of being "guilty of criminal fraud and murder."[52]

Radical dissent and the diversity of hypotheses it promoted was eventually limited to fringe AIDS communities, as consensus on the role of HIV in AIDS crystallized for both scientists and activists. Still, in the early days of the epidemic, confusion ruled. Within a context of homophobia—which most gays had to endure personally to achieve self-respect and the respect of society—many basic public health prevention measures were not taken lightly by the gay community. To those communities, such municipal efforts—which included the closing of gay bathhouses—seemed like an attempt to destroy what little had been achieved. The preventive screening of gay blood donors, carried out at a time when the cause for the disease was still unclear, was not well regarded either; given the backdrop of homophobia, such screening seemed to discriminate against gays.[53] The exhortations of their own peers to reduce their number of sexual encounters, limit the exchange of body fluids, or stop having sex altogether until they knew more about the epidemic (as some physicians had suggested) was rejected as evidence of internalized homophobia and fear of sex in general.

Those who spoke out against promiscuity—for example, Michael Callen and Larry Kramer—were equated with Moral Majority preachers who blamed gays for their own health problems.[54] The gay community rejected the bearers of bad news and anyone who publicized the less glamorous aspects of the gay lifestyle. This rejection was represented in the book *And the Band Played On,* by Randy Shilts,[55] published a few years into the epidemic (1987), which depicted the early years of AIDS denial within the gay community. As the epidemic continued to ravage American gays and medical evidence about its viral etiology and modes of transmission mounted, denial was replaced by acknowledgment that action should be taken. The changes in styles of activism have been documented in numerous accounts.[56] These accounts (mostly focused in New York City) provide the main sources for the history of AIDS activism in this chapter.

AIDS activism in New York can be divided into three main periods:

(1) The early period, which includes the initial responses and the creation of the Gay Men's Health Crisis (GMHC) and the People With AIDS Coalition (PWAC), related to the emergence of the PWA (Person With AIDS) identity.
(2) The peak period, during which the AIDS Coalition to Unleash Power (ACT UP) developed and Community Research Initiatives (CRIs) were implemented.
(3) The period marked by a decline in street activism and by the growth of establishment-based interventions; negotiations with governmental agencies, scientists, and pharmaceutical companies; the replacement of the CRIs with GMHC-associated CRIAs (Com-

munity Research Initiatives on AIDS); the formation of TAG (Treatment and Action Groups); and the shrinking of ACT UP due to the loss of many of its members to AIDS, collective burnout, and various political disagreements.

A fourth, post-activism period would follow, in 1996, corresponding to the impact of the new drug cocktail, which included the protease inhibitors. These drugs, as effective as they proved to be, appeared at the time of the total collapse of street activism and the growth of bureaucratized management of the chronic aspects of AIDS, as if coinciding with the routinization of charismatic reactions.

* * *

Even though this chapter focuses on activism in New York City, a few remarks about the social responses to AIDS in other sites should be included. The West Coast, for instance, was the other epicenter of the epidemic in the United States. In California, the cities of San Francisco and Los Angeles had parallel histories of responses to AIDS and created their own institutions. Some of these institutions resembled the ones in New York; such was the case with San Francisco's Project Inform, which was established as early as 1985 to compensate for the apparently minimal response to AIDS, the slow and inefficient pace of AIDS-related medical research, and what seemed at the time to be the overly cautious and slow FDA process for approving and releasing new drugs.[57] Other programs that were started in California to fill in the information gap regarding AIDS include the AIDS Foundation, the Healing Alternatives Foundation, and the County Community Consortium of the Bay Area HIV Health Care Providers (CCC). The Shanti Project, which had previously served people dying of cancer, also began to contribute to that effort.[58] Also, since 1987, the widely circulated newsletter *AIDS Treatment News* has kept the community apprised of developments in the field of AIDS-related clinical research.

Other urban centers soon followed suit. Activists in Boston, Washington, D.C., Philadelphia, Chicago, Toronto, and other cities founded their own ACT UP chapters, PWA associations, and AIDS service organizations. A "second wave" of activism hit other parts of the country a few years later, as the epidemic reached them, some time after it had struck the major urban centers.

Activists in European countries and Australia had their own versions of AIDS activism and developed their own community-based service organizations.[59] Among the most visible were the Terrence

Higgins Trust and Body Positive in London, AIDES in Paris, and the Deutsche AIDS Hilfe in Berlin along with its counterparts, the Swiss AIDS Hilfe and the Osterrëich AIDS Hilfe. These were large-scale organizations that not only provided a number of local services but also gave expert assistance to smaller organizations abroad. It is no coincidence that these types of services appeared where there had been some experience with democracy, civil rights, economic prosperity, and some expression of gay rights.

The developing world (see Chapters 3 and 4) had a completely different experience fighting AIDS. The absence of democracy and the often decades-long persistence of military or civilian dictatorships prevented the type of activism defined by a demand for individual and non-mainstream rights. Most local activism was directed by more basic and general social goals. In the developing world, AIDS activism grew in strict cooperation with development agencies, because they injected the money, models, infrastructure, and technical know-how needed to support local efforts against the epidemic. In most African and Asian countries, it is almost impossible to trace any connections between AIDS action and local gay activism, because the latter was virtually nonexistent. However, local AIDS activism was influenced by international gay activism, not only through its impact on central agencies, but also through the presence of foreign gay activists engaged in the development of global action.

This is also true for Latin America, which combined several components of AIDS action with pre-existing political activism in a way that deserves further analysis. In Brazil, for example, even though there had been some gay activism in urban areas during the late 1970s[60] (as expressed in the newsletter *Lampião*[61] and carried out by the group SomoS),[62] most of the responses to AIDS were not explicitly related to that movement.[63] Instead, AIDS activists adopted the language of international development agencies and combined their efforts with local social activism and the quest for human rights. It was not until the later stages of AIDS activism that it was connected to the gay movement, and in a very peculiar way: through the sponsorship of the government and the World Bank aid programs.[64]

* * *

In New York City, the earliest responses to the new epidemic preceded the coining of the term "AIDS." Raising funds for research on Kaposi's sarcoma, holding meetings to discuss "AID,"[65] and publishing articles in the *New York Native* (then the most popular and practically the only gay publication)[66] may have been the earliest expressions of public reaction to the epidemic. These early gatherings led to the creation, in 1982, of the first and what was to become the most prominent AIDS service organization in the world: the Gay Men's Health Crisis (GMHC).[67] This organization benefited from pre-existing networks within the settled, middle-class gay communities of Manhattan and Fire Island rather than from the more politicized veins of the gay movement. According to Randy Shilts, there was resistance to asking political leaders to join the action against AIDS, for they were seen as "downers" by the crowd that was at greatest risk. Instead, leadership was given to the more glamorous and attractive club types. Directorship of GMHC, for instance, was first given to the handsome yet semi-closeted corporate executive Paul Popham, and not to the out-of-the-closet writer just turned activist Larry Kramer, who had created the agency almost single-handedly and organized its first fund-raising event in his own living room.

GMHC started with volunteers, doing its own fund-raising and recruiting additional volunteers through the personal networks of the initial core group. People who were motivated to volunteer were those who were seeing their friends and lovers die, who felt sick themselves, or who believed they were at risk. Little was known about the new disease and how to prevent it. The federal and city governments were not doing anything about it, either. The authorities did not seem to be interested in getting involved with a "gay" problem, so the gay community had to take the initiative to do something. GMHC's initial team did its best to create awareness among the gay population and to encourage the community to change lifestyle-associated behaviors that might be causing the new disease.[68] There was no consensus around such issues, however. Opinions ranged from avoiding sex altogether to completely denying the link between sex and AIDS. As more evidence about HIV and its transmission came to light, lines of consensus came about, and safe-sex guidelines were established.

There was also a significant amount of discussion about how to care for the ill and dying. GMHC developed programs to assist people who were ill, including those in the terminal stage of the disease. They developed a quite successful "buddy system," through which a volunteer was assigned to a sick person to give him personal, psychological, and practical support. This program was crucial, because many people in the city did not have any relatives or healthy friends to rely on for the most basic help. The sick often needed more than a volunteer buddy could provide, however. Whether the agency should offer more complete home care or demand it from the city and state became an issue of disagreement within the GMHC. This tension did not destroy the agency, but it created a schism between those who favored providing assistance directly and those who preferred to demand that the government provide the necessary services.[69]

In spite of its less political position, the agency had strong support from the community and reached out to form more and better links with it. GMHC created a popular information hotline and developed extensive outreach programs, which were initially targeted to gay men and later expanded to reach other populations. Research within the community and conducted within the terms of the subject, "reaching out for subjugated knowledge,"[70] became the basis for the development of prevention programs. These programs included a variety of methods of disseminating information in order to promote safer behaviors, including workshops about safe sex and how to communicate with sexual partners, outreach programs that provided free clean needles or condoms, and programs on relapse prevention. Another sector in which GMHC developed its expertise from scratch was HIV-test counseling, which included pre-test counseling, test-based decision making, and post-test counseling. According to the agency's former director of education:

> GMHC's success developing efficient, effective prevention programs lies in the fact that our educational efforts are rooted in the values, experience, and language of the people we serve. (Reinfeld 1994:180)

The fact that the community was central to the design and language of their programs allowed the GMHC to obtain the type of data and develop the types of techniques that "experts" in medical science or public health could not, no matter how good their intentions might have been.[71] By focusing on the insiders' knowledge and attempting to incorporate it into every action, this agency was at the forefront of developing a new method of producing knowledge. This was one of the legacies of the responses to AIDS: the institutionalization of a new format for producing knowledge, one that grew out of the interaction between the community and the experts.

GMHC later became a major institution with a permanent staff, large amounts of federal funding, a sophisticated headquarters building in a fashionable neighborhood in Manhattan's West 20s, and a relatively complex institutional structure. Under its prominent institutional persona, it was criticized right and left. The explicit sexual references and imagery in the preventive educational materials shocked conservative forces. A diatribe by Senator Jesse Helms (the conservative Republican senator from North Carolina) against funding AIDS prevention programs, for instance, was based on his distaste for a GMHC leaflet targeting gay men. His opposition to federal funding to fight AIDS precipitated a long history of antagonism between the gay community and the senator's supporters.[72] Radical activists—including GMHC founder Larry Kramer (who left the group after one year)—were disdainful of the agency's plan to work within the establishment, like "a bunch of Florence Nightingales,"[73] instead of questioning establishment policies.

The GMHC became a landmark institution—receiving visits from representatives of many states and countries, creating models for providing services to people sick with AIDS, and developing strategies for education in prevention and support. Meanwhile, a different AIDS-related social movement developed through a bicoastal effort. The People With AIDS Coalition emerged in Denver from a meeting of several people living with AIDS who were tired of being "patients" or "clients" of agencies (i.e., GMHC), and having other people make decisions for them.[74] The most vocal among them were Michael Callen[75] from New York and Bobbi Campbell[76] from San Francisco, who were central to the creation of the Denver Principles and the new resistance identity of People With AIDS (PWAs). The concept was a refusal "of the label 'victims,' which implies a defeat," and a departure from the concept of being a permanent patient, "which implies passivity, helplessness, and dependence upon the care of others."[77]

PWAs felt that they were the most knowledgeable about their own disease. Thus they were able to make a new and unique type of contribution to medical practice and research. They also helped to detach, at least nominally, the response to AIDS from its strictly gay identity, thereby allowing people of different backgrounds to join the movement. In the years to come, GMHC would also expand its clientele and, in spite of keeping its original name, begin serving women, minorities, and other clients whose primary identification was not that of being gay.[78]

Several analysts of the PWA movement define it as a tributary of the feminist health movement.[79] The motto of the women's liberation movement—"The personal is political"—and the book *Our Bodies, Ourselves*[80] inspired self-empowerment and liberation from patriarchal medicine. While medical knowledge defined AIDS as a terminal disease, people with AIDS (PWAs) defined themselves as people who were living with AIDS rather than passively waiting to die of it a few months after the diagnosis. The impact of this movement included advances in treatment, improvement in quality of life, and the development of preventive education developments. The momentum of this movement echoed in many other places; in Brazil, for instance, people adopted the description of *pessoa vivendo com AIDS* (person living with AIDS) rather than the medical term *aidético,* which carried the stigma of having a terminal disease.

Those who were part of the PWA movement had a love/hate relationship with the medical establishment. They rejected the prediction of their imminent death—which had been made so pompously by the medical establishment—by living longer. They did not reject mainstream medicine entirely, as did the New Age devotees who rejected pharmaceutical drugs in favor of meditation and assumedly less toxic herbs.[81] In contrast, PWAs were willing to take highly toxic drugs that had no proven efficiency. Some did so illegally; for example, PWAs in California crossed the border into Mexico to buy drugs that were not sold in the United States. Others pushed for clinical trials of experimental drugs, often volunteering as study subjects; for example, PWAs in New York fought for official trials of the Israeli drug AL 721.[82] In New York, the People With AIDS Health Group (PWAHG) was the first association to coordinate and dispense information about drugs that were hard to obtain.[83] PWAs were not anti-medicine: they just wanted the doctors to do more, produce more, and know more.[84]

* * *

A second period of AIDS activism in New York, beginning in 1987, coincided with the development of the Community Research Initiatives (CRIs) and the founding of the AIDS Coalition to Unleash Power (ACT UP).

Before the CRIs were created, a series of "guerrilla trials"—community-based research on drugs that might be effective against AIDS-related diseases—had been held, the first one coordinated in 1984 by Joseph Sonnabend, M.D., in New York.[85] Those trials constituted a spontaneous response by community-based physicians and PWAs to the apparently slow and inefficient procedures used in the United States to approve and release AIDS drugs. Later, they served as an arena for the investigation of a more diverse group of anti-AIDS approaches, which eventually constituted an alternative to the hegemony of antiretroviral therapy, which was supported by leading research agencies, including the National Institutes of Health (NIH) and pharmaceutical corporations.

In a way, the PWA movement anticipated the possibility of a therapeutic paradigm that departed from the single-agent, "magic bullet" model inherited from the early days of bacteriology.

Somehow (to raise now a topic that will be developed later on), it is possible to argue that AIDS action developed more than just prevention and therapeutic knowledge: the community and their physicians advanced the possibility of a new paradigm for thinking and acting on infectious disease. While the scientific community works and thinks in accordance with the dominant paradigm in immunology and infectious disease, which is one of war, military actions and counteractions, defenses, and bullets,[86] the community, without theorizing, developed the possibility of an alternative paradigm—one that did not develop full imagery, symbols, and language, but that underlay attempts to remain alive in an environment of multiple infections and to stay healthy while overcoming and avoiding them, giving the immune system more credit than that of a warfare device.

But germ theory was what oriented mainstream research. Like Rifampicin for tuberculosis, Salvarsan for syphilis, the polio vaccine, and the antibiotics that defeat most bacterial infections, some HIV-specific "magic bullet" cure for AIDS should be developed. However, the optimism that followed the discovery of a viral agent did not have a corresponding timely response; there were years of attempts to create the right "magic bullets," and ultimately it was a set of "coordinated shots" that provided the acceptable medical approach, in 1996. This magic-bullet approach to AIDS resulted in the development of the first anti-HIV drug: AZT. Initially, researchers were skeptical of the idea of attacking viruses directly. Because viruses, which are nonliving entities, survive almost like parasites within living cells, it was believed that a "bullet" marked for the virus would also destroy healthy cells. However, another antiviral agent—acyclovir—had already been used successfully to control herpesvirus infections, which had become quite widespread in adults during the 1970s. The success of this antiviral agent (marketed by Burroughs-Wellcome under the brand name Zovirax) made some researchers more optimistic about the potential for antiretroviral therapy.

Burroughs-Wellcome also helped the NIH develop AZT. It later manufactured and marketed the drug, taking credit for and profiting from its development.[87] Manufactured commercially under the name Retrovir, AZT was praised and welcomed by patients and doctors—as well as the stock market.[88] Even before its efficacy was proven, activists demanded its release. They equated the delay in its release (which was due to the FDA's normal bureaucratic procedures for drug approval) with murder, because people were dying from AIDS every day. The community with AIDS felt that approval of the only promising drug available (no matter how mediocre its effects might be) should not be delayed. Volunteers entered clinical trials just to have access to what they believed was their only hope.[89] When Retrovir was finally approved in March 1987,[90] it was sold at a tremendously high price: The cost of one year of Retrovir therapy exceeded $10,000.[91] Like the FDA and later the NIH and New York's City Hall, Burroughs-Wellcome became the target of activists. Demonstrations against corporate greed were added to the portfolio of political activities.[92] After ACT UP became involved, the price of AZT fell—to a still unaffordable sum of $8,000 per year. In an article that appeared in the *New York Times* on August 28, 1989, the cost of AZT therapy was described as "inhuman" and "extraordinary," making it "the most expensive drug in history."

While mainstream researchers were using the most sophisticated, high-tech approach to the new medical problem by attempting to develop expensive retrovirus-specific drugs, CRIs were promoting more down-to-earth life-saving strategies that targeted the multiple infections that were actually killing people with AIDS—such as *Pneumocystis carinii* pneumonia (PCP)—and making plans to develop immune-system boosters. The New York CRI was trying out an apparently simple yet dramatically efficient approach: prevention of the secondary respiratory infections that killed most of the people with AIDS. Because of their efforts, physicians at the Memorial–Sloan-Kettering Cancer Center in New York and health care providers associated with the CCC (Community Council Consortium of HIV–AIDS Health Care Providers) in San Francisco started giving aerosolized Pentamidine to patients with PCP who were enrolled in clinical trials. This proved to be an efficient method of preventing pneumonia and saving lives.[93] The preventive approach was so effective that it influenced the management of persons with AIDS throughout the world. In June 1989, it passed the FDA approval process.[94]

Joseph Sonnabend, a physician with a practice in Greenwich Village, played a major role in the development of the preventive approach. His treatment protocol for individuals with AIDS included Bactrim as a prophylactic against PCP.[95] His decision to add prophylaxis to his anti-AIDS protocol was based on the findings of his own practice-based experiments and community-based epidemiologic survey, as well as his experiences in the treatment of many cases of sexually transmitted disease.[96] Furthermore, Sonnabend participated in British research on the role of interferon in the treatment of cancer during the 1970s. This frustrating experience may have been the root of his skepticism regarding the "magic bullet" approach to HIV.[97]

Sonnabend,[98] who often expressed his belief in a multifactorial cause for AIDS,[99] consistently searched for a clinically comprehensive approach that would prevent secondary infections from becoming lethal, rather than for a method of targeting HIV alone. However, HIV was what motivated most researchers—as well as the media, industry, stockholders, and people with AIDS; soon, this physician was marginalized by the mainstream AIDS establishment. In 1985, he was replaced by Dani Bolognesi as editor of *AIDS Research,* a position he had held since 1982. Even the name of the journal was changed—to *AIDS Research and Human Retroviruses,* to emphasize the role of the retrovirus in AIDS. In the same year, Sonnabend lost his position in the AIDS Medical Foundation (AMF), an organization that he and Dr. Mathilde Krim had helped create for the purpose of raising funds for community research. After he stepped down, the community-based AMF was expanded to become the highly visible organization now known as AmFAR (the American Foundation for AIDS Research).[100]

Those details, which seem irrelevant once an official history of scientific discovery is established, should be reviewed under the scope of the history of ideas and the sociology of scientific knowledge. Seen from the standpoint of today, the tale of developing scientific knowledge about AIDS, including its etiology and treatments, looks like the regular tale of enlightenment: some were wrong, some were right, and the latter finally brought light to everyone, proving what was real. However, there is more complexity in the process, and medical knowledge has always been a tension of competing perspectives—none of them more "real" than the other. The emphasis on the infectious agent is one part of the story; a comprehensive model, with interacting factors, is the other. Excessive emphasis on the single-agent version may lead to the disregard of complementing elements. In the case of AIDS, the multiplicity of views was not simply a matter for sterile academic discussion. Passionate points were made and often those who dismissed the emphasis on HIV were seen as irresponsible. Yet they had their point. Michael Callen, a patient of Dr. Sonnabend and a quite vocal long-term survivor, showed the empowering effect of the multifactorial hypothesis. Instead of accepting a prognosis of six months to live, he took control of his life by seeking ways to prevent secondary infections. He shocked the AIDS community by dismissing AZT, the drug whose release they had fought for so hard.[101] He then reached out to other long-term survivors to find some common denominator that could explain their longevity. Although he failed to find an objective factor, he recorded his efforts and produced a book containing a powerful testimonial to the importance of finding new ways of extending the lifespan of persons with AIDS.[102]

CRIs tested many potential treatments, most of which were proven ineffective and were discarded. Many of the "wonder drugs" that charmed the AIDS community—from AL 721[103] to Compound Q (an extract derived from the root of the Chinese cucumber)[104]—failed to produce any remarkable improvements. Furthermore, as death rates diminished for people with PCP, other lethal AIDS-related diseases emerged, including *Mycobacterium avium* intracellular (MAI). In their search for a means of controlling this "new" opportunistic infection, investigators eventually developed the drug Rifabutin, which reduced the risk for MAI by 50 percent[105] and extended the lifespan for people with AIDS accordingly. If nothing else, testing these two therapies—aerosol Pentamidine for PCP and Rifabutin for MAI—saved lives and spared suffering. Having passed through this community-based process of achieving knowledge, they were added to standard anti-AIDS protocols around the world. Today, most of the doctors who prescribe them and the patients who take them know little or nothing about the community-based efforts that resulted in the approval of these drugs.

The unconventional nature of the CRI research efforts did not spare the project from conventional financial problems. Short of funding and burdened with a huge debt, while criticized by early supporters for spending too much for too few results and for not publishing their results in credible medical journals, the CRI had to close its doors in 1991. It was replaced by the Community Research Initiatives on AIDS (CRIA), a similar project created and maintained as part of the GMHC.

* * *

At that point, and for a number of years, the epicenter of the basic indignation that drives citizens to political activism was located elsewhere: in the unmatchable AIDS Coalition to Unleash Power, known as ACT UP. Nowhere within the AIDS crisis was the fervor of reaction and the drive for action and change as intense as in this group, which generated a new social movement.[106]

ACT UP's most frequently reported myth of origin refers to an inflamed speech by the playwright and activist Larry Kramer at the Gay and Lesbian Community Center in New York's Greenwich Village in early 1987. Kramer was a founder and later a critic of the GMHC.

He started "preaching" to the gay community about the dangers of a sex-centered lifestyle years before the AIDS epidemic was identified. His book *Faggots,*[107] which depicted the sexual excesses of the fast-track urban gay life, had already angered many in the gay community. His message had been despised and ignored: nobody wanted to see the party end[108] or to see such a negative caricature of the community displayed before the general public.[109] After the epidemic was identified, Kramer's preaching intensified, and his message took on a different meaning. His play *Normal Heart*[110] was performed at the Public Theater in Greenwich Village. Loved and hated, depicted as the "Angry Man"[111] or as a "volcano of anger,"[112] he became a central voice in the AIDS movement.

On that evening in 1987, Kramer reminded the audience of his article "1,112 and Counting," which was originally published in the *New York Native* four years earlier.[113] At that time, 1,112 cases of AIDS had been reported in the United States. Four years later, 32,000 cases were reported—10,000 of them in New York City alone. He used material that appeared in that article to drive home key points about the epidemic, which were still true:

> At the rate we are going, you could be dead in less than five years. Two thirds of you will be dead in less than five years. If my speech tonight doesn't scare the shit out of you, we're in real trouble. If what you're hearing doesn't arouse your anger, fury, rage, and action, gay men will have no future on earth. How long does it take before you get angry and fight back? (Kramer 1989:128)

He called for leadership by trying to incite the masses to organize and do something. He denounced the ways of the GMHC, the government, the medical establishment, the slow pace of research. He called for the community to organize in order to lobby for access to drugs. He spoke out against the poor conditions in New York City hospitals. By stating, "We desperately need leadership,"[114] he incited people in the room to get organized and meet again. Two days later, about 300 people met in the same room—and ACT UP was born.

Others, including activist/commentator Maxine Wolfe, have pointed out that a single speech made by a single person cannot create a social movement.[115] Kramer had addressed the masses time and again to no avail. What made the difference in 1987 was that he made his speech at the right time and in the right place. Discrimination and homophobia had "turned illness into a political crisis."[116] The seeds of a new social movement had been fermenting for a while; signs of an incipient social upheaval in the gay community, fueled by indignation and anger about the epidemic, had been seen a year earlier.

The growth of this new social movement was a contrast to the previous period of stagnation. Since the years of GAA (Gay Alliance Activism), inspired by the Stonewall rebellion, there had not been much street activism in the gay community. Community action, epitomized by the GMHC, had consisted of fighting for legal rights and developing community services. The Gay and Lesbian Alliance Anti-Defamation (GLAAD) league was founded in 1985 specifically to fight the homophobic, racist, and sexist manner in which AIDS was reported; and it was said to have carried out "the first [political] activity in the gay community in fifteen years."[117] However, GLAAD soon developed a bureaucratic structure, one that could not keep up with the increase in community demands that accompanied the expansion of the epidemic. Anger and despair reached the boiling point. A street demonstration on the Fourth of July in 1986 attracted a crowd consisting of more than 5,000 gays, lesbians, and friends. GLAAD tried to keep the crowd under control as they marched in lower Manhattan, from Sheridan Square to Federal Plaza. However, the marchers continued to Battery Park—the starting point for ferries headed to the Statue of Liberty—at the tip of Manhattan Island. There they demonstrated before the one million tourists and locals who were in the area for the Bicentennial celebration.[118] This type of protest anticipated what would become the ACT UP style of street activism. [119]

This was already six years into the politically conservative administration of President Reagan, when federal inefficiency in promoting the search for a cure for AIDS was so blatant that it was hard not to blame the apparent lack of concern on homophobia. Had AIDS been considered a disease of heterosexuals, activists argued, then every public means of action would have been mobilized to spare people's lives and reduce their suffering. Some activists depicted the situation as genocide, a crime against humanity.[120] Presidents were called murderers. Reagan, later Bush, and even Clinton after an election campaign during which he promised so much, were attacked for not addressing the epidemic adequately. Reagan's silence in the midst of so many deaths was shocking; his lack of action contrasted with the readiness with which his Republican predecessor, President Ford, had prepared people for a swine flu epidemic that never materialized. The first Reagan-appointed AIDS commission included no AIDS experts. Furthermore, neither Reagan nor Bush was comfortable with "gay" issues; they condemned the gay lifestyle as "deviant," to match what they believed were the values of their constituencies. Kramer compared the situation to the Holocaust,[121] implying that a group-specific mass murder was taking place. The aura of collective endangerment helped raise the consciousness of the AIDS community, and a very politicized new movement began.

The graphic project "Silence = Death" began to appear on posters and bumper stickers in wide white sans-serif letters beneath a shiny pink triangle on a black background. The pink triangle on a black background evoked both gay pride and a past of discrimination. The logo was adopted by ACT UP in 1986 and remained a part of the movement's image for years to come.[122] Later, the California ACT UP chapter created a counterpart logo—Action = Life—with black letters on a white backdrop. Soon, every major ACT UP action was accompanied by a top-quality graphic project: "Seize Control of the FDA," "Storm the NIH," "Target City Hall." Keith Haring's artwork was adopted for AIDS awareness posters and T-shirts. Graphics, artwork, sound bytes, and well-prepared press releases enhanced the movement's visibility.[123]

ACT UP capitalized on a kind of energy that is hard to replicate or describe. It brought back the spirit of street activism. It brought together gays and lesbians who had been separated by different agendas for at least a decade. And it is no coincidence that the new "Queer Nation" was born out of ACT UP activity, bringing to the movement a new attitude, ideology—and style.[124] A weekly magazine, *Outweek,* served temporarily as the twin voice for ACT UP.

These novelties were accompanied by increasing politicization. ACT UP raised awareness about the social problems behind the health crisis, which were foreign to many of its members.[125] To keep from turning into a bureaucracy, ACT UP did not rely on centralization and delegation, except for specific and pragmatic purposes. Having an internal democracy and ordered anarchy meant long and often painful meetings, but guaranteed that the bureaucratic centralism that had corrupted other organizations would not be replicated. ACT UP was able to organize complex street actions, from Wall Street and City Hall in the crowded streets of downtown Manhattan to the FDA building in Washington, D.C. and the NIH headquarters in Bethesda, Maryland. Their preparation for street activism included teach-ins, civil disobedience training, and the formation of ad hoc affinity groups to resist potential violence on the part of the police. Teach-ins were translated into books and files containing information that no one else had compiled. This information was used to lobby and negotiate with the government.[126] ACT UP's consciousness-raising extended from its members to the public. It developed a new style of media-capture, using prepared press releases, successful sound bytes, and photogenic and telegenic demonstrations. It addressed (even if it did not solve) the problems of gender and minority representation on government committees and within the constituency in general.[127]

Finally, ACT UP developed the concept of AIDS activism as being different from gay activism. As Maxine Wolfe noted, even though the majority of ACT UP members were gay, the group inspired other patient-based social movements—including minority demonstrations and breast cancer activism. Since gay and AIDS activists had been inspired by the struggle of women and by African Americans for their own rights in the first place, the inspiration for political activism had come full circle.

The legacy of AIDS activism included exposure of the slow pace of enrolling volunteers in cancer drug trials, an issue that was taken up by the NIH:[128]

> Patient empowerment . . . shaped the policy response to AIDS and eventually inspired broader changes as advocates for people with Alzheimer's, breast cancer, schizophrenia and other diseases began to approach their diseases with new militancy. (Arno and Feiden 1992:61)

For more than five years, ACT UP attracted hundreds of people to its weekly Monday evening meetings. At one point, the crowd could not fit in the main room of the Community Center on West 13th Street, and meetings were held in the auditorium of the Cooper-Hewitt Building on Cooper Square. Meetings were an exercise in "structured anarchy," where anyone on the floor could cast a vote and no one had a greater right to speak than anyone else; in fact, no one was allowed to speak for more than a couple of minutes. During the meetings, attendees combined their intense sociability and exuberance with their ability to get things done. Several committees were established to provide the materials that were necessary to keep the crowd informed. Photocopies of relevant information gathered during the week—from press clippings to NIH or government bulletins—were also made available. The extensive amount of information provided by ACT UP committees guided the membership through the complexities of AIDS treatments and government politics and kept them abreast of ACT UP teach-ins and outreach and prevention efforts. Because of the revolutionary nature of ACT UP strategies, they became the subject of several books and articles of social analysis.[129]

Even within this egalitarian organization, tensions arose across the lines of gender and race more than once. Gay white boys were often accused of trying to dominate the committees; African American members often felt alienated from the group, even though some remained; Latinos developed their own caucus—ACT UP Americas—which developed links to South and Central American countries; women had to force their own agenda, which led to the publication of a book on women and AIDS.[130] The role of women in the group remained a polemical topic.[131] Political issues involving gender, race, and class were sometimes seen as distracting by members who wanted to focus only on AIDS. However, some of them realized that these issues had to be addressed in order to overcome the health crisis completely.

* * *

The decline of ACT UP and the emergence of specialized groups that negotiated within the establishment marked the third phase of AIDS activism in New York City. The anti-AIDS movement then split into different directions. One coincided with the agencies that provided services and struggled to keep their funding. A second one corresponded to the remaining grassroots activists—who no longer mobilized huge crowds but disdained both the AIDS industry and those who grew in AIDS activism and "sold their souls" to be insiders in Washington or sit on the advisory boards of drug companies. A third direction corresponded to those who specialized in drug research and protocol recommendations, like the Treatment and Action Group (TAG), and who were more prone to monitoring medical research than to demonstrations or community action.

TAG split from ACT UP in the fall of 1991,[132] taking with it a number of activists (such as Peter Staley and Mark Harrington) who had been prominent in ACT UP and in other community activities. TAG became a closed, by-invitation-only society—in great contrast to the inclusiveness of ACT UP, which allowed anyone who walked through the door to become a member with voting rights. The exclusiveness of the new organization angered some of the old-time activists. They spoke out against the careerism of the new "experts," who had traded their ACT UP T-shirts and baseball caps for corporate-looking suits and ties,[133] and had accordingly changed their working mindset.

Despite its controversial organizational structure, TAG was able to do quite notable work in monitoring high-level AIDS research. For the first time, activists shifted their focus from demanding immediate therapeutic success to a more comprehensive analysis of ongoing research—including basic research—and the production of medical knowledge. Perhaps they were capitalizing on the experiences of the AIDS community with AZT, which, a few years after its hasty release, was no longer seen as a wonder drug. The results of the Concorde trials,[134] which were made public in 1993, helped cast doubt on the efficacy and adequacy of administering the drug to asymptomatic HIV-positive individuals, as the standard procedure was then. Pharmaceutical companies and the NIH were sponsoring research on other nucleoside analogues, including didanosine (ddI) and stavudine (ddC), which, like AZT, target the enzyme reverse transcriptase. Although these drugs piqued the interest of the general public (and the stock market) temporarily, they did not offer a real breakthrough in AIDS therapy. Something else was needed, including a change in the line of scientific inquiry.

TAG helped the scientific community define new lines of inquiry that could lead to the development of more effective drugs. At the Eighth International Conference on AIDS, held in 1992 in Amsterdam, Mark Harrington presented a paper on HIV pathogenesis and activism at a plenary session attended by basic research scientists. Raising central questions concerning AIDS pathogenesis,[135] he argued that it was worthless for scientists to replicate nucleoside analogues such as AZT if they still did not know how HIV infected human cells. He called for more cooperation between the AIDS community and research scientists. He also called for more research on one of the more neglected strategies for fighting AIDS: immune system boosters, pointing out that while the mainstream scientific community focused on antiretrovirals and the community-based research initiatives focused on preventing secondary infections, the complexity of the immune system, particularly within the context of AIDS, was yet to be understood.

By specializing in the language of science, and, to a certain extent, replicating the scientific research mindset, TAG was able to make demands that were more in line with the way science is normally carried out than were those of other community advocates. A clear example of this arises in the discussion about the release of protease inhibitors (drugs that block the activity of enzymes that activate reverse transcriptase, the enzyme that results in the replication of HIV in infected cells). While AIDS activists had always advocated for the fast release of drugs, TAG asked for longer periods of investigation so that the mistakes that were made with AZT would not be repeated.[136] TAG also monitored ongoing AIDS research, looking for duplications and gaps, raising new questions to be explored, and providing recommendations to the NIH and pharmaceutical companies.

The work of ACT UP's special group on Treatment and Data was parallel to that of TAG. In 1992, the ACT UP subsidiary developed the Barbara McClintock[137] Project to analyze, supervise, and recommend questions for study and the means by which they could be approached. It eventually evolved into the "Cure for AIDS" Project, which focused its energy on promoting a cure for AIDS—anticipating, perhaps, news of the triple cocktail that was to be announced four years after the McClintock Project was initiated.

In 1995 ACT UP still held Monday evening meetings—though with fewer in attendance—back at the Gay Community Center. It held a number of media activities and appeared in collective public demonstrations, and promoted consistent work by the Treatment and Data Committee. For example, after Clinton's election in 1992, ACT UP staged a mock funeral in Washington[138] to express their displeasure with Clinton's failure to keep most of his campaign promises to support AIDS research and PWA programs, and to promote AIDS prevention. But the organization's energy was drained, with too many people lost to AIDS or to burnout.

The GMHC continued to expand. By 1995, it had a staff of 270 and 6,500 volunteers, with an annual budget of $27.5 million. Each year, it served more than 7,000 people directly, and tens of thousands more through its advocacy and education efforts. In fiscal year 1994, it distributed 750,000 pieces of literature on HIV/AIDS, and its hotline answered 46,000 phone calls. The gay white male profile of its original clientele changed; it was now 25.1 percent black and 26.8 percent Latino, with heterosexuals accounting for 28.4 percent of all clients, and women accounting for 16.4 percent.[139]

In a city with an estimated 200,000 people infected with HIV, where the number of cases reported in mid-1995 exceeded 75,000 and the number of adult deaths had mounted to 48,877, AIDS service organizations are constantly drained by demand and continue to multiply. New Yorkers became accustomed, sadly, to what had once been a novelty for the city. By the mid-1990s, many New Yorkers had lost close friends, neighbors, or family members to the epidemic. The entire city faced losses that had earlier been the almost exclusive experience of the gay community. Drained by losses and by the trivialization of the epidemic, grassroots movements ceased to shake the streets. Scientific AIDS research entered a phase of "normal," rather than "revolutionary," science, to use Kuhn's terminology. Annual international meetings became biennial. It was this exhausted society that received the news of a promising new therapy—a triple cocktail that included the promising protease inhibitors—in 1996. In a cover story that appeared in the Sunday Magazine section of the *New York Times,* journalist Andrew Sullivan expressed the optimistic view of the news in the article's title: "When Plagues End."[140] A number of other articles focused on the unexpected hope for people with AIDS who had been contemplating their own death and who could now look forward to living again.

* * *

The efforts of social movements can no longer be ignored in scientific AIDS research. Some of the achievements of such partnerships are enumerated by Bruce Nussbaum:

> The changes in trial design, the beginning of parallel track, the curbing of placebos, the use of surrogate markers instead of death in tests, the quick testing of drugs at the community level, are all revolutionary research initiatives. (Nussbaum 1990:332)

This author refers to such achievements as things that should have been done long ago. In his view, the reason why better knowledge about AIDS was not available earlier lies in the lack of accountability of today's medical researchers:

> ACT UP, the CRI, Project Inform, AmFAR have consistently shown over the recent years that people with the disease, their doctors, and their advocates know more about treatment than do ivory-tower principal investigators hidden away from the realities of life and driven by careers that don't reward them for furthering public health. (Nussbaum 1990:332)

Like other journalists covering the epidemic,[141] Nussbaum has an individualistic view of history, emphasizing the role of characters and personalities in the course of events. My approach is different. Even though some of the individuals who intervened in the history of the social movement against AIDS are mentioned in this book, I have tried to provide enough background to allow for a social interpretation of the events.

The shift in the social actors involved in the production of medical knowledge changed the direction of research; the meaning of this shift to a discussion on the social construction of knowledge will be examined in Chapter 6, after a number of wider elements have been presented for analysis.

THREE

Sponsoring Global Action

The Role of WHO

IN MY ANALYSIS OF AIDS ACTIVISM and its effect on the production of knowledge (Chapter 2), I argued that the characteristics of the social movement that responded to the AIDS epidemic in the United States were closely intertwined with the political culture of this society. That "political culture" stems from a number of traditions, including the fight for civil rights and the consequent expression of individual and group rights, the praise for individual freedom, the feminist health movement, the gay rights movement, and an emphasis on self- and community empowerment. Some of the experts involved in the development of transnational responses to AIDS, such as Jonathan Mann and his colleagues, ask whether patient-based activism on AIDS is a specifically U.S. phenomenon or whether it became legitimately international.[1]

Social responses to the AIDS epidemic developed everywhere, including regions whose sociopolitical characteristics diverged substantially from those of the United States. These responses were diverse and partially coherent with the local cultures where they developed. However, they should not be studied solely in reference to their own context. The transnational sphere, its agencies and programs, and the influence of U.S.-based activism throughout the world, were all elements that influenced each and every local response. The role of the World Health Organization (WHO) in sponsoring responses to AIDS around the world—particularly in the developing countries—deserves special consideration. This agency was itself influenced at some level by the U.S. social movement against AIDS and marked its special program with an emphasis on human rights and social issues.

This chapter will show the close interdependence of the local and global levels of action and decision making; subsequent chapters will take their perspective from a local setting in Brazil and will be followed by an analysis that also involves the field of biomedicine. For now, I will address the local/global interdependence, from the perspective of the global sponsorship of WHO. The understanding that AIDS affected the developing world had an impact on the agency's decisions and was followed by the implementation of a global action program.

In the context of AIDS, WHO promoted partnerships between developed and developing countries, governments and nongovernmental organizations, scientists and the public. Whether such global efforts and new partnerships actually provided the fermenting hybridization that was needed for innovative approaches to the epidemic (see Chapters 1 and 2) has come into question.

* * *

Although as recently as 1993 more cases of AIDS were identified in the United States than elsewhere in the world,[2] it had become increasingly clear that AIDS was not just an urban gay disease in developed countries. In fact, it had become a global pandemic,[3] varying in its effect from region to region and with a particularly harsh toll in the developing countries.[4] The evaluation of the global impact of AIDS reached apocalyptic levels of dismay and frustration.

In 1993, WHO estimated the number of new HIV infections at 5,000 a day.[5] By the year 2000, the number of projected infections ranged from 30 million to 40 million according to WHO,[6] or from 38 million to 110 million according to the Global AIDS Policies Coalition.[7] Despite disagreements over the quantitative methods used to calculate these projections,[8] both agencies stressed the catastrophic nature of the epidemic and its burden on the developing countries—which, according to WHO's estimates, would account for 90 percent of all HIV infections in the year 2000.

Projections were used to plan ahead and to raise awareness, to try to persuade all nations to develop or improve AIDS programs, to pressure their at-risk populations to adopt preventive behaviors, and to lobby donor agencies for funds to fight AIDS. The projections may have been overblown, but the figures at the time were quite disturbing. In 1993, 850,000 cases of AIDS were reported to WHO by more than 180 countries. Given the stigma of the disease and the difficulties of diagnosis in countries with high mortality rates, the agency believed that this was a low estimate—that the actual number of AIDS cases in the world that year might have been more than 3 million.[9] As for the number of HIV infections, whether they turned into full-blown reported or unreported AIDS cases or remained asymptomatic, the estimate for 1993 was about 15 million total cases of infected persons, of whom 1 million were children. About 12 million of the total number of cases of HIV infection corresponded to living people.[10]

The geographic distribution of the epidemic has been described as uneven—about 1 million infections in North America, 1 million in South America and the Caribbean, 500,000 in Europe, 30,000 in Australia, and 30,000 in North Africa and the Middle East. Sub-Saharan Africa had the largest share, estimated at 6.5 million. While East and Central Asia had an estimated 20,000 infections each, and Australia about 30,000, South and Southeast Asia had the highest estimated number, at 1 million HIV infections.[11] According to WHO, the rate of infection in Southeast Asia in 1993 paralleled the rates seen in Sub-Saharan Africa during the early 1980s.[12] The predictions for that region were grim unless efficient measures were taken.

Changing its approach to the epidemiological patterns of this disease, the Global AIDS Policies Coalition[13] developed the concept of "Geographic Areas of Affinity" (GAA), which divided the world into ten regions according to their shared characteristics.[14] The ten areas that were described by the "New Global Geography of AIDS" were North America, Western Europe, Oceania, Latin America, Sub-Saharan Africa, the Caribbean, Eastern Europe, the South and East Mediterranean, North and East Asia, and Southeast Asia (including India). Indices used by researchers to characterize the epidemic in each area included the estimated year in which HIV infections began to spread (1977 to 1978 for Sub-Saharan Africa, 1978 for North America and Western Europe, 1978 to 1979 for Latin America, 1979 for the Caribbean and Oceania, and 1982 to 1984 for the remaining areas); the year in which the first case of AIDS was diagnosed (the early 1980s for the Americas, Europe, the Caribbean, and Northeast Asia; the mid-1980s for Sub-Saharan Africa, Southeast Asia, and Oceania; and the late 1980s for the South and East Mediterranean); the availability of AIDS/HIV data; major modes of HIV transmission; the ratio of urban to rural cases; the seroprevalence of HIV; the male:female ratio (which ranges from 10:1 in Eastern Europe to 1:1 in Sub-Saharan Africa, with intermediate ratios of 4:1 in Latin America and 8:1 in North America); additional operational data, such as the first year of national response,[15] the level of external funding (which is low everywhere except in Africa and Southeast Asia) and the percentage of the country's gross national product (GNP) that is spent on health services; and a number of socioeconomic and developmental indicators that are used by the United Nations Development Programme (UNDP), such as the human development index, the total female score, the human freedom index, and the annual growth rate for the urban population.[16]

Using the Delphi method to calculate the number of infections in each geographic area, the Coalition came up with estimates that differed from those of WHO. According to the Coalition's Global Report, there were an accumulated total of 12,875,450 infected adults and children throughout the world,[17] with the largest number in Sub-Saharan Africa (8,770,000 infections, or 68 percent of the total), followed by North America (1,180,000 infections, or 9.2 percent of the total), Latin America (1,040,000, or 8.1 percent), Western Europe (730,000), Southeast Asia (700,000), and the Caribbean (330,000).[18] The Coalition also predicted that the infection would grow exponentially; however, the "pace" of growth differed from that predicted by WHO. Nonetheless, the two organizations shared a pessimistic outlook regarding the impact of AIDS on many aspects of society, especially demography, economy, and culture. The only positive outcome, many would acknowledge, had been the vitality and creativity of responses to the epidemic, particularly the creative methods used by those involved in the social movement against AIDS to influence the direction of clinical research and AIDS policies.

> A logical outcome of the successes of AIDS activism in the industrialized countries (as exemplified by . . . ACT UP), will be to connect issues and struggles in the developing and industrialized countries. Thus, while access to AZT was reduced by high prices in the industrialized world, suitable trials and access to AZT, other antiretroviral agents and drugs to treat opportunistic infections are all extremely limited or absent in the developing world. A history of AIDS activism would also include many courageous and creative efforts in communities and countries around the world. Women refusing to continue subservient sexual roles, community groups working for human rights and dignity, groups working to permit sex education in schools, and people with HIV and AIDS fighting discrimination—all are activists. Their collective story needs to be documented and told. It is a global story which is making global history. (Mann et al. 1992: 240)

Signs of a social breakthrough appeared in the midst of the global devastation caused by this epidemic. WHO tried to bring the world together. It mediated interactions between centers and peripheries, making a point of giving voice to the generally silenced peripheries. Within the realm of the global fight against AIDS, WHO and other international agencies helped to create a worldwide infrastructure for easier and affordable communications. Fax machines and modems, enabling easy access to electronic mail and bulletin board services, were made available almost everywhere. Travel funds were also made available, so that those who had been confined to the peripheries could become acquainted with the centers where decisions were made.

Their voices became more audible through a number of electronic and print media, as well as through conferences and meetings.

Our question is whether these efforts resulted in a significant change in center–periphery interactions, or whether they simply brought a new outlook to old hierarchies. The process of science-making is usually portrayed as a hierarchical polarity between the "First" and "Third" worlds, whether in a "modernization" or a "dependency" model (see Chapter 1). A more interactive scenario would require a radical departure from this order of things. Could a global scenario serve as the locus for a paradigm shift in the biological sciences (see Chapter 1)? Global interactions might result in as fertile a field for the creation of new knowledge as was the interaction between activists and the scientific agencies in the United States (see Chapter 2).

Brazil seemed to be the right place to investigate this problem because of the vitality of local medical research. Several fields with strong traditions of studying the social dimensions of medical problems (social epidemiology, infectious diseases/immunology, and the clinical practice in tropical medicine) could develop enough possibilities for creative and fertile interactions that might improve the extent of our knowledge of AIDS.

* * *

WHO's leadership in the international AIDS movement after 1985 and 1986 transformed the global response to the epidemic. This is documented in several newsletters and world reports, including the *Global AIDS News,* published by the Global Programme on AIDS (GPA); *Aids Health Promotion EXCHANGES,* which was originally edited jointly by the GPA and the Royal Tropical Institute of the Netherlands (KIT) and eventually by KIT and the Southern Africa AIDS Information Dissemination Service (SAfAIDS) in Harare, Zimbabwe; *AIDS Action,* published in several languages by the London-based Appropriate Health Resources and Technologies Action Group (AHRTAG); and collective volumes, such as the series edited by the Panos Institute;[19] the comprehensive *Global Report: AIDS in the World;*[20] or WHO's *AIDS: Images of the Epidemic.*[21] Also, many government documents on AIDS that were published throughout the world after 1988 were influenced by WHO, as they attempted to promote programs that followed the guidelines of the agency.

The circulation of AIDS information through these publications indicates a change in the universe of knowledge dissemination, if not in the process of creating knowledge. Did this change influence the world of science? In one respect, it did: it influenced the direction in which information flowed between the social actors concerned with AIDS. Traditionally, medical knowledge travels from research specialties to the clinical sphere and from there to the social sphere, not the other way around. Clinicians use the knowledge created by virologists, for example, and social workers use the knowledge created by clinicians. AIDS newsletters permitted information to circulate in the opposite direction, making available to biological researchers the knowledge of community-based organizations.

Also, the format adopted by the main forums for developing knowledge on AIDS had a special, interactive character. The International Conferences on AIDS were truly interdisciplinary, bringing together different disciplines for intense dialogues. The participants included people with AIDS (PWAs), community organizers, and a significant number of representatives of developing countries.[22] The format of the conferences did not replicate that of traditional medical forums, in which interdisciplinary dialogue rarely extended beyond contact between basic researchers and clinicians. In contrast, the AIDS conferences included clinicians, basic researchers, epidemiologists, social workers, social scientists, activists of various sorts, and representatives of the pharmaceutical industry, governments, nonprofit organizations, and traditional healers—a rainbow of diversity that sometimes made the traditional scientists nostalgic for the single-discipline forums. The large AIDS conferences were different from any that had existed before, because they incorporated the double impact of AIDS activism and global sponsorship.

Another question concerns the effect of the global fight against AIDS on the traditional hierarchies within these disciplines. By giving particular attention to the voices of the Third World, WHO had an effect on transnational scientific communities. Traditionally, the degree to which a community is heard is directly proportional to the status of its local scientific community within the international system. Central, industrialized countries are considered to be the producers of "legitimate" scientific knowledge. Peripheral countries merely consume such knowledge or, at best, adapt it for local use, but they are not "assigned" to produce mainstream knowledge. A similar hierarchy generally exists between patient and doctor. Through innovations brought about by the fight against AIDS, however, the previously "silent" patient inverted this hierarchy of knowledge and power (see Chapter 2).

Was anything similar happening in the center–periphery hierarchy, in terms of "making science"? Did the global action against AIDS reverse that hierarchy to allow research ideas and findings to flow from developing countries into the mainstream? Were the new international partnerships that had been forged by the fight against AIDS to become the foundations for the cross-fertilization that might produce breakthrough research? And if so, would they reverse the traditionally top–down, North–South, center–periphery direction of the dissemination of knowledge? Was the developing world participating actively in a process that was important for the world in general, and most of all for the developing world?

Following the rhetoric of WHO and the information disseminated through many international forums and newsletters, one might think that the global response to AIDS had resulted in such a transformation. If that were true, the structures that fostered dependency might be in the process of being replaced by dialogical interactions and global communication. I risk saying that the experience of participating in the global fight against AIDS, particularly in 1989 and 1990, gave many a taste of that utopian possibility.[23] That was also a time of "utopian treatment activism" in the United States,[24] which coincided with the end of the Cold War and was imbued with a social optimism that is hard to match. The world seemed to be undergoing enormous transformations. The visible effects of the new partnerships and global action could be seen in the shape of social responses and in the development of community programs.[25]

* * *

The First International Conference on AIDS, which was held in 1985 in Atlanta, Georgia, was not your average scientific forum. At that time, AIDS was a mysterious disease. The virus known as LAV/HTLV-III (see Chapter 1) had been identified less than a year before, and relevant information had changed so rapidly that a thorough review of all available information was sorely needed.[26] For that purpose, more than 2,000 people from fifty countries met in Atlanta.[27] The event has been repeated every year, with growing participation and visibility.[28]

By 1985, it was already evident that the epidemic was far more than an "urban gay disease." Right from the beginning, in the United States AIDS evoked male homosexuality, but from a European-based international perspective, AIDS evoked Africa rather than male homosexuality. "Africa" was not then the symbol of exoticism and remote origins it became later in other contexts; it represented the harsh reality of Africans getting sick in rapidly growing numbers. Many of the European institutes of tropical medicine regularly handled cases of severe infectious diseases brought from former colonies in Africa by migrants and visitors. Most of the first AIDS patients either were African or had been in Africa. The African link seemed plausible to French researchers involved with the initial identification of LAV. American reporter Randy Shilts describes AIDS in Europe as consisting of two epidemics, the first one linked to Africa and the second one linked to American gays.[29] Back then, however, the idea of AIDS as an African disease was as odd to Americans as its being a gay disease was to most Europeans or Africans.[30]

The AIDS epidemic had apparently been devastating large parts of the continent for longer than it had been affecting American gay communities. The characteristics of epidemic clusters[31] in Africa were different from those in the United States.[32] It took several years for the epidemiology of African AIDS to be defined in its own terms—that is, around clusters of roads, traffic, warfare, and other social variables rather than on the basis of individual behavior, as it was in the United States. Thus, in contrast to being a predominantly "homosexual" disease, African AIDS was defined as predominantly "heterosexual." This variable, like the other, is linked to individual behavior and characteristics, as opposed to the social variables that were later used to account for the epidemic in Africa: multiple deprivation, accumulated STDs, poverty, economic depression, continuous warfare, migrations, and even iatrogenic causes.[33]

The fact that American AIDS had been characterized as "homosexual" and African AIDS as "heterosexual" illustrates how epidemiology draws on local social constructions and how the global characterization of the epidemic was asymmetrical and one-sided, treating a particular local knowledge—in this case, of the United States during the early 1980s—as universal.

AIDS might not have started in America,[34] but it was there that it had its "cognitive birth," a fact that both spared time and lives[35] and forever marked the understanding of the epidemic. Later, a search for its origins helped make it visible everywhere, including in developing countries, where symptoms of AIDS blended with many pre-"epidemiological transition" illnesses and premature deaths. The retrospective, North–South tracking of the epidemic globalized and apparently unified the world under a common problem: we all shared the threat of the new epidemic. This unification was only partial, however. Its epidemiology was based on individual behavior (homosexual vs. heterosexual) and it perpetuated a dual (I/II) pattern model that only emphasized the schism between developed and developing nations. The reference for defining the epidemic pattern of AIDS in any country—in Brazil in particular, as we will see in the next chapter—became that of an "American" epidemic or an "African" one, as if doubling the opposition of development and underdevelopment.

The diversity of patterns was framed by WHO reports as basically consisting of an opposition between pattern I and pattern II.[36] Pattern I corresponded to the developed countries, and was characterized primarily by infections contracted by homosexual transmission, plus some intravenous drug transmission. Pattern II was seen in Africa and the Caribbean and was characterized mainly by heterosexual transmission.[37] Pattern III corresponded to the countries with few reports of infection. The contrast clearly mirrored the opposition between nations that were characterized in terms of post-epidemiological transition and those that were in pre-epidemiological transition. Such reporting—which was still primarily bipolar—strongly influenced the thinking of epidemiologists, clinicians, basic researchers, activists, and policy makers throughout the world for many years.

Interestingly enough, it was in Brazil that this partitioning of a world epidemic was criticized. The Brazilian political writer and later AIDS activist Herbert Daniel challenged the use of established biomedical and epidemiological models to characterize AIDS. Many other public writers and social scientists,[38] most of whom were connected to the Associação Brasileira Interdisciplinar de AIDS (the Brazilian Interdisciplinary AIDS Association, or ABIA) (see Chapter 4), soon followed suit. The influence of ABIA on the nongovernmental agencies fighting AIDS (AIDS–NGO) movement in Brazil, as well as the influence of activists such as Daniel and the others involved with the emerging social movement, partly explains the resistance of Brazilian activists to the traditional epidemiological categories. Time and again, newsletters from Brazilian NGOs challenged their government's use of epidemiologic "imported models"[39] to characterize AIDS in Brazil. Brazil had a specific epidemiology, they argued, the terms of which should be a priority for research. To see it through "imported models" and "imported categories" would only create an artificial understanding of the epidemic in Brazil—something that activists constantly accused the government and the media of doing.

In order to expose the mechanisms that resulted in such a distorted understanding of the epidemic, a group of researchers that included two anthropologists and a political activist deconstructed the standard categories applied by doctors reporting AIDS cases. Their empirical data, which were derived from the files of the first 200 persons reported with AIDS in Rio de Janeiro,[40] led them to conclude that the official epidemiological categories were inadequate for local cultures[41] and actually helped conceal the true face of the epidemic in Brazil.[42] They suggested that the characteristics of AIDS that were specifically related to the epidemic in Brazil should be researched first, a task that ABIA made a priority on its agenda (see Chapter 4).

At the same time that Brazilians were challenging traditional methods of characterizing AIDS, the anthropologist Richard Parker[43] argued that Brazilian sexual ways could not be reduced to the limited categories used by international agencies, which were based on contemporary U.S. culture and focused on the distinct "opposition" of homosexuality and heterosexuality.[44] Other anthropologists later argued similar points for other Latin American sexual cultures.[45]

Parker argued that those points were extremely important when it came to developing AIDS prevention policies. Rather than presuming that sexual culture is the same everywhere, local customs and categories should be investigated and should form the basis for the development of any local AIDS policy. Brazil was illustrative of an argument that should be valid for every culture in the world. Parker's work had a significant influence in global agencies,[46] and the author became an international broker who often served as the official voice of the social-behavioral component of AIDS research within global networks.

Local Brazilian efforts to revise the categories and epidemiological patterns used to characterize AIDS had a wider counterpart in *Global Report: AIDS in the World*,[47] a volume that is dedicated to the memory of Herbert Daniel.

* * *

WHO cherished the global concept and applied it to health problems long before it became a buzzword in sociology during the late 1980s and 1990[48] and before it became a pet expression for corporate business.[49] The agency's introductory publication, *Fighting Disease*, described its take on mass disease as "a global approach to global problems."[50] The global problems discussed in that publication were tuberculosis and malaria, which still ravaged Third World populations while European and North American populations were enjoying a period of expansion, economic growth, and social development. The program created later to develop research in tropical disease was conceived as a "global partnership."[51]

Like other UN agencies, WHO was created in the spirit of optimism and cooperation that characterized the post–World War II era.[52] Global by definition and inheriting the Office International d'Hygiene Publique,[53] WHO's first three decades were marked by the optimism for growth and progress that characterized the 1950s, '60s, and '70s. Its success in the eradication of smallpox, another "global program,"[54] helped consolidate the idea that it might be possible to defeat disease worldwide. Social scientist J. Peabody[55] argues that the success against smallpox increased the pressure on WHO to eradicate other ills that ravaged the Third World. Medical writer Laurie Garrett points out the role of that success in the decline of Infectious Disease as a medical specialty in the developed world.[56] And yet, as this author shows later, "victory" was an illusion and far from permanent.[57]

* * *

Under the one nation–one vote World Health Assembly, WHO's main constituency was clearly that of the developing countries, even though most of the agency's funding came from the few developed countries in the United Nations. Post-war economic development did not generate equal or even a proportional distribution of benefits for all, but rather increased the webs of dependency and unequal development. It was within this economic context that Third World health problems became a primary target for the agency. Africa was somehow the very embodiment of the Third World, its problems aggravated by the effects of recent and late colonialism. Responding to African crises of famine or drought and their health implications, the *WHO Chronicle* reported that:

> the crisis that Africa continues to face is not purely one of recurrent drought or famine; it is a development crisis affecting all fields of endeavor, including food and other rural production incentives, social planning and gradual alleviation of the unfavorable socioeconomic conditions that prevail throughout most [of] Africa. (WHO 1986c:234)

WHO originally defined health not only as the absence of disease but also as general physical, mental, and social well-being. Given this definition, it would take more than just handing out medicine to kill microbes and vaccinations to prevent them to ensure good health around the world. It would take, in the agency's words, global efforts and global strategies. In the 1970s, these efforts were crystallized in the project Health for All by the Year 2000[58]—which now seems so overly optimistic, given current trends in AIDS, tuberculosis, malnutrition, infant mortality, and the quality of drinking water around the world.[59] In his organizational analysis of WHO, Peabody describes the Health for All goal as "vague," "utopian," and "unreachable."[60]

Following the 30th World Health Assembly (1977) and the International Conference on Primary Health Care in Alma-Ata, in the former Soviet Union (1978), the principles of the Health for All project became the driving force of the agency. In 1981, the Global Strategy for Health for All by the Year 2000 was adopted by the Health Assembly.[61] "Global" seemed to be the agency's favorite concept: there were "global strategies" and "a global policy for health,"[62] as well as "global promotion and support,"[63] "global information exchange,"[64] "global orientation and support for research," "global use of national expertise,"[65] "global monitoring and evaluation,"[66] "global level," "global coordination," "global enlistment,"[67] "global promotion and coordination of research and development,"[68] "global advisory committee on medical research,"[69] "global partnership,"[70] and "global policy framework."[71] Given the international scope of AIDS, this epidemic became a major concern for WHO. The agency's "earnest involvement"[72] with the new epidemic began with the First International Conference on AIDS in 1985 in Atlanta, the WHO Consultation on AIDS, and the EURO Consultation on AIDS.[73] In October of that year, WHO organized the first meetings on AIDS in Africa, to be held in Bangui, Central African Republic. Besides Central Africans and WHO officers, there were participants from Rwanda, Burundi, Uganda, Tanzania, Zaire, Cameroon, and Gabon. The purpose of the meeting was to establish a cooperative data-gathering effort in deprived areas.[74] In December 1985, the American branch of the Pan American Health Organization (PAHO) created the working group on AIDS Guidelines. By the end of the year, another WHO conference on AIDS was held in New Delhi, India.

A few other AIDS-related scientific meetings were sponsored that same year: the WHO informal meeting on T-lymphotropic Retroviruses of Non-Human Primates (Geneva, July 1985); the Scientific Group on Non-A–Non-B Hepatitis, Delta Antigen-Associated Hepatitis[75] and Blood-borne Human Retroviruses (Tokyo, October 1985); the Consultation on AIDS–Safety of Blood and Blood Products (Geneva, December 1985); and the Second Meeting on Simian Retroviruses (Geneva, December 1985). In addition, two meetings of the "WHO Collaborating Centres on AIDS" were held in September and December 1985.

The existing collaborating centers and the related networks constituted the infrastructure on which WHO could rapidly build its global program on AIDS and a system for monitoring the epidemic as well as responding to it globally. The initial program included:

> the exchange of information, the preparation and distribution of guidelines and educational material, support for laboratory diagnosis, cooperation in the development of national programs and action of AIDS containment, advice on the provision of safe blood and blood products, and the coordination of research. (WHO 1986b:52)

Even though it had already been working informally, the Special Programme on AIDS was officially established on February 1, 1987, invested with the "responsibility for the urgent mobilization of national and international efforts and resources for global AIDS prevention and control."[76] The AIDS epidemic was defined as a "world health problem of extraordinary scale and extreme urgency," representing "an unprecedented challenge to the public health services of the world."[77]

The Global AIDS Strategy was endorsed unanimously by the World Health Assembly in May 1987,[78] presented to the Venice Summit of Heads of State and Government in June of that year,[79] and formally adopted by the United Nations General Assembly in September.[80] In January 1988, it was presented to the World Summit of Ministers of Health and Programmes for AIDS Prevention in London, resulting in an early compilation of papers.[81]

The Special Programme on AIDS, and later the GPA, made a point of involving the developing countries, especially those in Africa, in the fight against AIDS. The fact that by 1985 more than 90 percent of the reported AIDS cases had come from the United States and Europe[82] was not interpreted by the agency at face value, but as hiding the asymmetries in systems of reporting:

> Neither the overall number of cases reported nor their countries of origin present a true picture of the epidemiology of AIDS, since there are vast differences between countries in funds and other sources available for the surveillance of disease. (WHO 1985b:211)

For our analysis, the lines that follow are the most interesting. In them, WHO states its hope that a collaboration with the developing countries will make a significant contribution to the advancement of knowledge about the epidemic:

> It is our hope that the mechanism established by WHO for international surveillance will yield a more complete picture of AIDS and hasten the progress towards solving its more profound mysteries. (WHO 1985b:211)

These statements seem to imply that there were expectations about a knowledge breakthrough and that the Third World had an important role in the process.[83] Were the expectations limited to such areas as community action or maybe epidemiology, or did they extend to the clinical and basic sciences? The field with the most intense and visible new interactions and multidirectional flows of information was the area of prevention, which counted on the involvement of local organizations. This was also the area that received the most attention in the newsletters mentioned above.[84]

A significant amount of the funding and expertise provided by WHO was channeled not only into the implementation of national programs on AIDS but also into community-based organizations. Too little time and too many lives were at stake in this complex epidemic to overlook any possible action—no matter how unconventional. With the impact of the AIDS epidemic, the entire Health for All by the Year 2000 project was bound to collapse prematurely. Following the successful examples of American community-based strategies to fight AIDS, WHO tried to involve, fund, and create community-based organizations throughout the developing world in order to fight off the epidemic. They endowed these community-based organizations with the technological means to interact constantly with international agencies in order to share information quickly and efficiently. At that time, fax machines, microcomputers, CD-ROMs, and electronic networks became available. Many of these technical devices were exploited in the development of AIDS communication and networks.[85]

Under the leadership of Jonathan Mann and a few tireless and sensitive health experts, the GPA became a dynamic and innovative agency within WHO.[86] Mann became quite popular in Third World countries and among community activists; he encouraged the involvement of the community, respect for human rights, and compassion for people with AIDS. The top-level agency and its representative had learned from the grassroots level that a problem such as AIDS should not be addressed with the strict public health measures used for other epidemics, for example, quarantine and contact tracing to separate the sick from the healthy. Such actions would only deepen the alienation of those at risk, force the epidemic underground, increase the number of infections, and worsen the already difficult situation of people living with AIDS. It seemed that the agency had learned something new and different from the traditional and hard-line public health style.[87] The involvement and engagement of people with AIDS and the community organizations that served them became an official policy of WHO, and as such was passed on to donor agencies and local governments.

The guidelines provided by WHO for the development of national AIDS programs in different countries[88] were far more progressive than any that most governments (many of them being part of authoritarian regimes) would ever have developed. The GPA tried to make governments understand the importance of developing national programs and funding them. WHO helped them get the funding, either by lobbying for it or by providing it directly. Some of the recommendations were hard to implement with funding alone, for cultural and ideological elements created as powerful an obstacle as the lack of resources. For example, the guidelines spoke candidly about drug abuse and homosexuality,[89] whose existence many governments refused to acknowledge and always considered a foreign phenomenon. As late as 1985, Japanese medical authorities dismissed the local relevance of AIDS, because they believed there were no homosexuals in Japan.[90] As late as 1988, Chinese authorities claimed that there were no AIDS cases in China, because homosexuality, promiscuity, and drug use were prohibited by Chinese law.[91]

Also, amid the dry language of medical policy, the guidelines included the recommendation to "promote the use of condoms."[92] In countries where the Catholic Church held sway, this was hardly conceivable, much less feasible.[93] The GPA tried to persuade governments to include people with AIDS and community-based organizations in the design and implementation of these actions. This concept seemed extraordinary to public health officers, whose guidelines for epidemics consisted of isolating the sick and protecting the healthy from contamination. Also, many activists, speaking as people with AIDS, were opponents of ruling authoritarian regimes. For many governments, the idea of including "revolutionary" PWAs in the planning of official anti-AIDS policies was almost unthinkable. Yet, GPA had discovered the potential for this formula by following the actions of patient-empowering movements, especially those in the United States.

Between the publication of the Guidelines for the Development of a National AIDS Programme[94] and the publication of guiding principles for monitoring them (a period of just one year),[95] the role described for PWAs in WHO manuals changed substantially. By 1988, they were referred to in a small entry as deserving care and compassion:

> Because of the psychological and other effects of HIV, those most directly affected (persons already infected, with or without clinical illness), along with their sexual partners, household members, and others in the environ-

> ment, must be helped with their problems through counseling, education, and other ways. (WHO 1988c:7)

One year later, PWAs were presented as active and necessary partners in the fight against AIDS:

> Their own experience makes individuals with AIDS potentially excellent educators about HIV infection and its consequences. They may be particularly effective as sources of information about aspects of AIDS that is difficult to convey otherwise or for target audiences that are not easy to reach through conventional information and education programs. (WHO 1989a:37)

WHO probably reached the peak of its globalization efforts in the fight against AIDS between 1989 and 1990. Connections with Third World countries and activist organizations were a priority for the GPA. NGOs throughout the world praised Mann's leadership and contributed to the GPA's innovative energy. In countries such as Brazil, where both NGOs and the government had strong links with the GPA, the restructuring of the Programme in late 1990 (which included Jonathan Mann's replacement by Michael Merson) was initially seen as unfortunate.[96] Even though the new office maintained the support for NGOs and social action, enthusiasm for the novelty of community participation gave way to a more medicalized and epidemiological approach.

In the following years, there was some competition between GPA and the newly formed Global AIDS Policies Coalition for global leadership in the acquisition of social knowledge and promotion of international social action against the epidemic. The Coalition was now coordinated by Jonathan Mann and a number of collaborators based at Harvard, and with partnerships with many different countries.[97] Its visibility and influence peaked between 1991 and 1992 with two major global contributions: the Eighth International Conference on AIDS, organized in 1992 in Amsterdam as an alternative to Boston because of the continued restrictions on HIV-positive travelers to the U.S.;[98] and the publication of the mammoth report *AIDS in the World.*[99] This book was rapidly translated into Portuguese and published in Brazil,[100] where the Coalition maintained a significant influence (see Chapter 4).[101]

In January 1996, GPA was replaced by a larger interagency office in the United Nations—UNAIDS, headed by Peter Piot, who had been associated with AIDS action since the very early years of the epidemic. The creation of an interdepartmental agency had been demanded for a long time by many of the donors, experts, and activists involved with AIDS throughout the world.[102] In order to act on some important fronts, the global program needed the cooperation of agencies such as the World Bank, the Food and Agriculture Organization (FAO), the United Nations Children's Fund (UNICEF), and the United Nation's Development Program (UNDP), which had been peers rather than subordinates of WHO.

As of this writing, WHO's role in the fight against AIDS is at the end of a cycle. Some analysts have already ventured an evaluation of that role, which has naturally involved a number of problems. According to Peabody, WHO treated AIDS within a more or less successful framework that had a number of problems, which he describes as the "yaws approach,"[103] taken as the example of WHO's organizational culture. The "yaws model" was named after the successful 1950s campaign against yaws, a disfiguring infectious disease caused by an agent similar to that of syphilis. That campaign involved international meetings, the training of national staff, and the distribution of technology,[104] and served as the prototype for other programs, consisting of "technical meetings, consultative visits, fellowship training, and the provision of supplies."[105] In this author's analysis, the yaws model is inadequate for diseases such as AIDS that do not have "magic-bullet" cures. In such cases, other approaches should be explored.[106]

In my interpretation, however, there was more to WHO's action against AIDS than the mechanistic "yaws approach," particularly during the first years of the GPA. The GPA had made a point of including the experiences of people living with AIDS and their contributions to the AIDS struggle in their program-planning efforts. It also facilitated global awareness and the development of a global approach to AIDS that reduced the risk of arbitrary local action. The staff of the GPA worked hard to achieve that goal. Mann and his team traveled constantly to give the problem visibility and to show the complexity of the approach required: not a simple medical issue, but one that included its social and political dimensions. In so doing, they created a communications network to enhance the global nature of the approach to AIDS and to develop a network of partnerships that could move quickly, communicate easily, and subvert traditional hierarchies between countries and between disciplines.

Although WHO developed a few global programs for worldwide health problems, it did not keep up with the innovative direction that the fight against AIDS had taken. As author Laurie Garrett notes, the GPA shook the bureaucratic structures in WHO's headquarters in Geneva and raised obstacles within the organization. The emphasis by GPA on human rights as a key issue in the fight against AIDS generated many negative reactions.[107] The fact that its statistics had smaller figures for AIDS than for the traditional infectious diseases, such as TB or malaria, was used to argue against the special attention to the new epidemic. Hiroshi Nakajima, director of WHO from 1990 until this writing, tried to get the GPA to follow the agency protocol; however, GPA officers had already established their own innovative style. Their split from WHO resulted in divided loyalties throughout the world, and new approaches to the AIDS epidemic—which had raised in so many people and in so many places a type of consciousness that had never been raised before—were replaced with a more predictable and less mobilizing medicalized model. However, the structures and awareness created during the most active years of the GPA had influenced the actions and ideas of many throughout the world. The year 1990 may have marked the end of an era for relying on total dialogue, the subversion of hierarchies, the development of creative partnerships, and the existence of multiple revolutions. But hope for a general transformation was in the air, and the possibility of change was more than an intellectual exercise: it was the key to keeping alive for just too many people.

FOUR

Local Action

Responding to AIDS in Brazil

IN MOST PLACES OUTSIDE THE UNITED STATES, AIDS became an issue before any local cases were reported. Because AIDS was often perceived as a foreign disease, it inspired an ideology that was somehow independent of its actual course. In some cases, it was considered an artificial, genocidal device, like a cold war weapon that was out of control.[1] For example, in Africa, which had been blamed as the birthplace of HIV, AIDS was thought of as a form of sabotage against the Third World.[2]

So distinct did this ideology become that Brazilian writer and PWA Herbert Daniel used the term "aids," written purposefully in lower case, to refer to the ideological and political phenomenon, "a significance that means far more than the disease referred by the acronym AIDS."[3] The same distinction was followed by activist Silvia Ramos, who wrote,

> aids arrived to Brazil before the first cases of AIDS. An epidemic (the third) [of prejudice] starts to circulate in Brazilian society with a large impact right in the beginning of the early 1980s. The expectations and excitement of the national soul were bigger than the effects caused when the disease arrived, grew among the population and, now, explodes . . . a few years later. (Ramos 1990:8)

The intricate relationship between global action and local responses to the AIDS epidemic can be explored through the example of Brazil. The epidemiology of AIDS in Brazil is of the same kind as that in the United States,[4] but the political cultures of the two countries are quite different. Brazil's political culture should be part of our analysis, due to its impact in the local forms of activism, clinical practices, and representations of epidemiology.[5]

Like most of Latin America, Brazil was subjected to Catholic Iberian colonialism beginning in the sixteenth century. Its economy depended largely on gold mining and on the slave-based plantation system, the root of social stratification that remains in large parts of the country. Social stratification is more visible in the sugar cane–producing areas of the northeast, from Pernambuco to Bahia, although it marks the entire country, including its older capital, Rio de Janeiro. The southern states (including São Paulo), which were settled more recently by entrepreneurial European migrants who arrived after the country became independent (1822), are less stratified than the northern states, and their residents enjoy a higher economic status.

When Rio de Janeiro became the capital of the Portuguese colonial empire in 1808, as a strategy to avoid the republican wave that swept Europe after the French Revolution in 1789, Brazilian political independence became irreversible and a fact after 1822; Rio would forever retain the attitude of a "center," even though it ceased to be the capital of Brazil in 1960, when the administration was moved to Brasília. Brazilian independence did not prevent the country from remaining in a tight economic dependency in a complex colonial and post-colonial world system. Universal citizenship and democracy remained fragile and often interrupted projects.

After decades of monarchy, attempts to create democratic structures were brief interspaces within larger periods of republican populism and, as in other Latin American countries, military dictatorships. Under such circumstances, it is not surprising that the North American style of civil rights and the notion of self-empowerment are quite remote in Brazil—at least for the majority of the population. This element will emerge from the analysis of responses to AIDS in Brazil. While the fight for individual rights became a central agenda in the United States, Brazil was under a military dictatorship that lasted until the early 1980s. Only after that did the fight for civil rights and citizenship take shape, becoming a mark of the new social movements in Brazil.

That was also the moment when the AIDS epidemic struck, and AIDS gained visibility within the new social movements. In several instances, the two agendas merged: civil rights were not a given, but something that had to be fought for. In some cases the energy of AIDS activism helped strengthen the awareness of and fight for civil rights. The response to the AIDS crisis in Brazil also coincided with internationally sponsored efforts to develop a global response to AIDS. Sponsors included the World Health Organization, several donor agencies from the developed countries, and coalitions of people from developed and developing nations. Funding increased as the global response to AIDS increased, enabling organizations that were coordinating international efforts to help shape local responses to AIDS.

In most countries with Latin-based languages, AIDS became *SIDA* (for Syndrome de l'Immuno-Déficience Acquise in French, Sindroma de la Imuno-Deficiencia Aquirida, in Spanish, or Síndroma de Imunodeficiência Adquirida, in Portuguese). In Brazil, however, the English term AIDS was retained, although with a unique pronunciation: "i-DEEZH." In the first series of government manuals on the treatment and prevention of the epidemic, it was referred to as SIDA/AIDS,[6] and in clinical settings it was often referred to as SIDA, as in other Portuguese-speaking countries. Among government agents, AIDS activists, health professionals, and in the media, however, AIDS was used as a noun rather than as an acronym.[7]

In Brazil, as in many other settings outside the United States, AIDS became newsworthy for the simple reason that it came from the United States,[8] and even more so because this mysterious disease seemed to affect mostly famous people and gays, who were considered exotic. Morbid speculations about the disease were raised in the Brazilian media—Was it going to come to Brazil? Was Brazil "sophisticated" enough for such a disease to spread? In a country burdened with a split identity—developed yet underdeveloped—it is not uncommon for the media to explore the ambiguities of collective identity. If Brazil were to have AIDS, it would mean that on some level Brazil was a developed country; after all, AIDS was a First World disease. As a consequence, some of the early cases of AIDS attracted the spotlight, in contrast to the shameful silence given to other endemic diseases that affected millions of poor Brazilians, reports of which were not considered fit to print. Readers did not want to know about the 8 million people with Chagas disease, the 13 million with schistosomiasis, or the 460,000 new cases of malaria that were reported in 1987 alone.[9]

AIDS, in contrast, deserved media attention as a *Doença de Primeiro Mundo* (disease of the First World) and was popularly referred to as a *doença de rico* (disease of the rich).[10] As Moraes and Carrara[11] note, the symbolic opposition between the Third World and the "civilized world" was the most complex one within a repertoire of images that were used by the media to portray the epidemic, and which included references to gender, sexuality, and death. Coming from the "primitive heart of Africa" to "slay" New York, the "capital of the world," the new epidemic was portrayed as the "revenge of the oppressed." Passing from monkeys to Africans to Cubans and Haitians, the virus was said to be transmitted to American gays on vacation in those countries. Ranging from primitiveness to modern decadence, an array of symbols loaded accounts of the spread of this disease.[12] The epidemic in Brazil was portrayed in the media as an encounter between the "civilized" and "primitive" extremes, the authors note:

> It gets delirious when the presence of the disease among us is seen as a sign of distinction and civilization. When [the magazine] *Manchete* publishes in the first page the title, "AIDS: Brazil is already the vice champion," it is not just a morbid joke: there is also a certain amount of pride in there. AIDS becomes the proof that among us exists the lifestyle of the big American metropolises, that we are also "civilized," that the "sexual revolution" happened here also. (Moraes and Carrara 1985a:17)

The representation of AIDS as an affliction of the rich remained solid for many years because of media reports that focused almost exclusively on celebrities with AIDS. We now know that the epidemic probably affected several social strata equally from the time it first reached the country,[13] although this possibility remained unexplored. That the incidence of AIDS among the poor deserved media attention[14] and public awareness came about only after the relationship between AIDS and poverty attracted media coverage in the United States. In 1991, "pauperization" and "Africanization" became buzzwords used by the media to describe the epidemic in Brazil. The patterns I and II defined by the World Health Organization (WHO) (see Chapter 3) formed the basis of these classifications. Furthermore, when some women became sick, the media declared that the epidemic had changed from pattern I to pattern II, thus essentially suggesting that the condition had been "Africanized." The concept of Africanization of AIDS in Brazil—which had already been mentioned during an interview that was reported in 1987 in the daily *O Globo*[15]—was often attributed to international agencies such as WHO[16] or PAHO.[17] This line of thinking led to misinformation and the reinforcement of certain stereotypes. In an article that appeared in O *Globo* on April 13, 1989, HIV-2 was linked with stereotypes associated with poverty, such as undernourishment, poor hygiene, and promiscuity, which were declared:

> the ideal conditions for the explosion of AIDS, through a change in the patterns of transmission: instead of the propagation by risk groups (homosexuals, injecting drug abusers), today dominant all over the world except in Africa, the passage to the heterosexual transmission. (*O Globo,* April 13, 1989)

Giving in to a racial connotation that was rarely stated explicitly, the article further suggested that it was concern about the Africanization of AIDS in Brazil that had led the research agency FIOCRUZ to send its leading AIDS scientist, Bernardo Galvão, to Bahia to create a research laboratory for surveillance of the epidemic. The rationale was that "the area, besides having a large number of people of African descent, has the basic conditions for the propagation of the virus in the heterosexual population" (*O Globo,* April 13, 1989).

Virtually every researcher who I interviewed dismissed the notion of "Africanization," on the basis of both ideology and biology. In fact, the variants of HIV found in Brazil belong to the same "clade," or group of variations, as those found in the United States and Western Europe, that is, HIV-1B. Also, the dominant mode of transmission was sexual—mainly from man to man, with some reports of man-to-woman transmission and fewer cases of woman-to-man transmission.[18]

* * *

The association of AIDS with privilege at the beginning of the epidemic in Brazil had two repercussions. On one hand, it was used as an excuse for the government's lack of attention to the disease,[19] as the government could claim that AIDS affected a very small minority (the "First World within"), who could afford their own health care abroad, and that it should use its resources for the massive endemic diseases, such as malaria and tuberculosis, that were affecting the deprived in massive numbers. In a page-length article that appeared in the daily *O Globo* on September, 8, 1985, Health Minister Carlos Santana declared that even though his office was taking a series of measures against AIDS, this disease could not really be considered a priority. There had been only 415 confirmed cases of AIDS, when there were 4.5 million cases of Chagas disease, 7 million cases of schistosomiasis, 385,000 of malaria, 206,000 of leprosy, and 88,000 of tuberculosis—not to mention the 1.5 million cases of sexually transmitted diseases (STDs), such as syphilis and blenorragia. On the other hand, the publicity brought to AIDS by celebrities helped mobilize solidarity campaigns and fund raising. After some pop stars and TV idols became sick or died,[20] support from the performing arts community for the AIDS cause in Brazil increased significantly.

The first publicized case of AIDS in Brazil reinforced its stereotype as a disease of the rich and famous. The disease struck São Paulo's well-known gay designer Markito in 1983. Markito was seen as a cosmopolitan gay traveler who was believed to have contracted AIDS in New York. He was also portrayed as the irresponsible type who went back to the metropolis for fun and partying instead of getting treatment.[21]

Cosmopolitanism contributed at least as much as homosexuality to the initial symbolic imprint of the new epidemic, and Markito's case was used as evidence that Brazil shared the international jet-set characteristics associated with AIDS. Contrary to the homophobic censorship that prevented the early publication of information about AIDS in the United States, Brazil made a show of this disease. While in the United States the rampant number of cases could hardly get any coverage, in Brazil the few initial cases were the subject of excessive, morbid attention.[22]

A second basis for public awareness of AIDS in Brazil, and one that mobilized a more political type of attention and action, was the diagnosis of AIDS in Henfil, the well-known and beloved political cartoonist who had criticized and fought the military dictatorship with his art. Henfil had hemophilia and was exposed to HIV through iatrogenically contaminated blood, as were his brothers Chico Mário, a musician, and Betinho (Herbert de Souza), a politically active sociologist who was widely known as a community organizer.[23] Betinho helped shape the perception of the epidemic as a public issue that involved empowerment, citizenship, social compromise, government policies, social control, and the fight for basic human rights. The issues of illegal blood banks, corruption, and government inefficiency were made visible through commentary on Henfil's disease and helped mobilize organizations with wider agendas.

The fight for citizenship in a country that was emerging from an authoritarian military rule (1964 to 1985) provided a larger and better umbrella for anti-AIDS organizing than a gay-based movement might have provided, many activists argued.[24] Besides, there were no distinct and established gay communities[25] in Brazilian cities as there were in New York and San Francisco. Moreover, there is a pervasive ideology that denies the existence of homophobia in Brazil, similar to the denial of racism—which, naturally, takes other and more insidious forms.[26] If there is no homophobia, the argument goes, then it makes no sense to construct a gay movement or gay-based political actions.

The question of the existence of homophobia is more complex, however. Scholars have portrayed the subtleties of Brazilian sexual culture in a number of ways, describing everything from a "closeted" gay life within a repressive homophobic culture[27] in which original categories such as *entendidos* develop,[28] or a place where an incipient gay movement struggles[29] to bring about a shift from hierarchical to egalitarian relationships,[30] to a *sui generis* sexual culture in which the categories of "gay" and "straight" are subverted by erratic erotic meanings such as *sacanagem,* which corresponds to transgression rather than to any clear definition.[31] There are also detailed ethnographic descriptions of a clandestine urban gay life,[32] historical interpretations of same-sex interactions in past centuries,[33] and cultural interpretations of sexual meanings.[34]

In spite of the diversity of interpretations about the characteristics of Brazilian sexual culture, a gay movement arose within the context of the Abertura (literally, "Opening," the movement toward democratization within the military dictatorship of the late 1970s), which corresponded mainly to SomoS and Lampião, but it did not last long.[35] When the AIDS epidemic began in Brazil, only a few of those groups remained. Even though they addressed the problem and developed AIDS-related projects, such as that of the Atobá in Rio de Janeiro and the Grupo Gay da Bahia (GGB) in the city of Salvador, Bahia, these groups did not provide the main leadership in the fight against AIDS. Most of the structured responses to AIDS were developed, rather, under the banner of human rights, in particular by nongovernmental organizations (NGOs).

* * *

ONG, which stands for Organização Não-Governamental—(literally, nongovernmental organization, or NGO), became the most popular acronym for the community-based organizations involved in the fight against AIDS in Brazil. Rather than being supported by local gay organizations, Brazilian AIDS activism was largely connected to the international mobilization efforts against the epidemic. The very format ONG/NGO indicates some level of international support and external aid to Brazilian AIDS activism. The link between such organizations and WHO-sponsored forums would only increase in the following years and expand to other fields.[36]

Like community organizations throughout the world,[37] NGOs were central players in Brazilian social movements against AIDS during the decade from 1985 to 1995. They played an important role in the production of knowledge about the epidemic and developed extensive networks that facilitated their interaction with other intervening social actors such as international agencies, the Brazilian government, and the medical establishment. Unlike community organizations in the United States (see Chapter 2), Brazilian NGOs were defined not only in terms of their relationships with the government, the medical establishment, or other local social forces, nor only by the local dynamics of people with AIDS; they were also defined through strong connections with external international social forces, both through funding and through the sharing of knowledge.

It can be said that international awareness of AIDS began in 1985 (see Chapter 3). That was the year of the First International Conference on AIDS and of the launching of WHO's special program. It was also the year of Rock Hudson's death from AIDS and its scrutiny by the American media.

Public awareness also started at that time in Brazil. *Cadernos ISER,* the journal of the Higher Institute for the Study of Religion (ISER),[38] dedicated its seventeenth issue to AIDS.[39] A group of São Paulo's citizens—including health care professionals and social activists, some of whom had experience in the gay movements within the *Abertura*—created the first and largest Grupo de Apoio à Prevenção à AIDS (AIDS Prevention Action Group), or *GAPA*. The following year, the largest Brazilian AIDS agency—the Associação Brasileira Interdisciplinar de AIDS (Brazilian Interdisciplinary AIDS Association), or ABIA—was born in Rio de Janeiro. ABIA's founders included medical and social scientists, physicians, social activists, and a number of citizens concerned with the poor response of the Brazilian government to the AIDS epidemic.

During the following decade, a number of similar but smaller organizations were founded throughout the country; these types of organizations came to be associated with the social response to AIDS. In 1989, there were 51 AIDS–NGOs, including those that added AIDS to their pre-existing agendas.[40] There were 60 NGOs a year later,[41]and 87 in 1992.[42] In 1994, a government-published catalogue listed 140 such groups.[43] Twenty-two organizations claimed to have begun working with AIDS as early as 1986, 11 in 1987, and 8 in 1988. The number of organizations claiming to have started working with AIDS increased during the following years—20 in 1989, 18 in 1990 and 1991, 25 in 1992, and 17 in 1993.[44]

ONG was an uncommon term that gained popularity during the 1980s;[45] it neared exhaustion by the 1990s in several fields of social intervention in Brazil, Latin America,[46] and worldwide. Regarded as an "imported" term, "coined in the Northern countries,"[47] a term that had been "adopted for convenience"[48] from the jargon of international aid agencies, and consolidated during the Abertura, ONG became irreplaceable in the Brazilian repertory of social institutions. Its "myth of origin" refers to the 1950s and to religious activism.[49] Indeed, the Catholic grassroots movement, which was most visible in CEBs (base ecclesiastic communities), provided important background experience for many who became involved in the NGOs, including those with AIDS support services. Left-wing politics served as another source of organizational skills.[50] It was in the late 1960s, when the dictatorship was particularly harsh, that this organizational format matured, often as the possible legal face of political action.[51] A third source for the NGOs was the university. Nonprofit agencies act like magnets for "organic intellectuals." They are appealing work sites for those who want to avoid the sterility of the academic ivory tower, the slow bureaucracy of public service, or the stress and materialism of corporate businesses. In NGOs, university graduates could fulfill their social commitments by providing "intellectual support for the popular movements,"[52] either on a full-time or a part-time basis, and either out of strong political faith or just because they needed a job.

The social response to AIDS in Brazil coincided with the period of expansion and consolidation for NGOs. In the late 1980s, the acronym ONG occupied the symbolic spot formerly occupied by *movimentos* (movements), *associações* (associations), and *grupos* (groups) in the field of AIDS as well as other spheres of social intervention. Also, the term *ativista,* tailored after the international icon "AIDS activist," replaced the older term *militante,* which had been used by the political left wing. Thus, a new social vocabulary had emerged in response to AIDS.

AIDS–NGOs, like NGOs in general, varied greatly in form and function. Some resembled business offices that were housed in mansions and employed several full-time employees, used technologically advanced equipment, carried out a number of projects simultaneously, maintained multiple international connections, and received generous renewable grants. Others operated out of someone's back room, employed volunteers, and were funded by raffles, button sales, and small donations. Some were committed to political intervention, while others were devoted to assistance work. Some acted as monitors or critical partners of the government, while others developed their focus of activity from one day to the next.

NGOs implemented research projects, organized conferences, and published competitive pieces of work, or they engaged directly in social intervention without ideological, political, or scientific guidance. Such diversity is sociologically recurrent: "If heterogeneity is the rule within and among Latin American countries, it is gospel among NGOs and social movements."[53]

In his 1992 social profile of Brazilian AIDS–NGOs, Nelson Solano from GAPA–São Paulo presented them as a diverse group that shared common goals. His survey included 87 organizations, of which 51 were considered autonomous, 19 had religious affiliations, and 11 were linked to sexual emancipation groups. They were seen as a part of a new and spontaneous social phenomenon, considered by the analyst to be mostly autonomous (except for the religiously affiliated groups).[54] The author fails to account for the driving force of funding; it was funding that helped create, sustain, and sometimes define the work of these organizations. As NGO analysts Ruben César Fernandes and Leilah Landim commented, NGOs do not like to refer to their sources of funding and take it quite matter-of-factly.[55]

Solano compared the AIDS–NGOs with previous Brazilian social movements that pretended to represent the "popular classes." The new movement did not have that pretension, nor did it necessarily oppose the government—only its health policies. The AIDS–NGOs seemed to relate to urban centers with a relevant epidemiology or with clearly defined homosexual or religious communities, which are implicitly acknowledged as constituting the other driving force for this social movement. The author blames amateurism, a lack of reflection, and excessive pragmatism, as well as the corresponding absence of political, educational, and social goals and the absence of articulations with other sectors of society, for the identity crisis and other difficulties experienced by the movement.[56]

There were several attempts to organize the Brazilian AIDS–NGOs under a common structure, often inspired or supported by international networks. Among the many GAPAs that were created in the meantime in various Brazilian cities (a total of eighteen in 1992),[57] there were a few channels of communication and cooperation, such as through the exchange and borrowing of materials, ideas, and styles, and the sharing of experiences. Between 1987 and 1989 alone, there were five nationwide inter-GAPAs meetings.[58]

Even though much of the constituency of AIDS–NGOs was gay, they did not follow the pattern of gay-based AIDS activism in the United States. Rather, they followed global efforts promoted by international funding agencies. The AIDS–NGOs movements in Brazil experienced a turning point with the official acknowledgment of their role in the global fight against AIDS by the World Health Assembly in May 1989.[59] This acknowledgment followed the attempts of the Global Programme on AIDS (GPA) to formally include the community-based organizations in the fight against the epidemic.[60] This move helped empower, strengthen, and fund local groups throughout the world.

The highest moment in this movement was probably the Forum called Opportunities for Solidarity, which was sponsored by WHO in June 1989, in Montreal, prior to the Fifth International Conference on AIDS. More than 300 organizations and 600 activists from around the world participated and exulted in the acknowledgment of their transformative power. Some had already been incorporated into the inner planning and executive core; Brazil's ABIA, for example, helped organize the meeting.[61]

The language of networking was a key aspect of the Montreal meetings. The International Council of AIDS Service Organizations (ICASO) and the organization that was to become the Latin American Network of Solidarity were cited as examples of structures that were integrated for local action. The forum "inspired the Brazilian NGOs that participated to seek regional and global forms of participation and response."[62] Such forms were to be found by "the articulation through national networks, as a first step to the formation of regional and international networks, interconnected through ICASO [the International Council of AIDS Service Organizations]."[63]

NGOs had intense moments of empowerment and stood out among other groups involved in the fight against AIDS, such as those involved in medicine, science, and public administration. Several Brazilian organizations were present in Montreal. From Rio de Janeiro alone, there were ABIA, GAPA–RJ, the gay group Atobá, Prostituição e Direitos Civis (Prostitution and Civil Rights), and Ação Religiosa Contra a AIDS (ARCA), the last two being linked with ISER (the Higher Institute for the Study of Religion, a large umbrella NGO with many different social research and intervention programs); the GAPAs of São Paulo and Bahia, GETAIDS, from Brasília, and the Pernambucan gay group Movimento Antônio Peixoto were also in attendance.[64] They all returned home enthusiastic about creating a national network and executing common projects.

Shortly thereafter, the first meetings of the AIDS–NGOs Brazilian Network of Solidarity were held in Belo Horizonte, Minas Gerais (June 14 to 16), attracting fourteen organizations: the GAPAs from São Paulo, Rio de Janeiro, Bahia, Belo Horizonte, Pernambuco, Rio Grande do Sul, and Santa Catarina; ABIA, ARCA/ISER, and Prostituição e Direitos Civis/ISER; the recently formed Pela VIDDA (an association of people living with AIDS and their families and friends, inspired by Herbert Daniel); the group Solidariedade, from Belo Horizonte; the Lambda support center from São Paulo; and the young national "names"/quilt project, Projeto Nomes.

A committee formed at those meetings that included Nelson Solano (GAPA–SP), Jane Galvão (ARCA/ISER), Ranulfo Cardoso (ABIA), and J. Eduardo Gonçalves (GAPA–RS) was given the role of expanding the network and raising funds for the next meetings. They counted fifty-one NGOs and obtained international funds from the British agencies CAFOD (Catholic Fund for Overseas Development) and the Save the Children Fund, the U.S.-based Ford Foundation, and national donations from a local industry and a private donor.[65] Thirty-eight NGOs participated at the meetings held in Porto Alegre. There were already thirteen GAPAs present. In addition to those represented at previous meetings were the GAPAs of Baixada Santista, Ribeirão Preto, São José dos Campos, and Taubaté (all in the state of São Paulo), plus Belém (Pará) and Fortaleza (Ceará), as well as some groups with similar names, such as GEAPA, GETAIDS, and GEPASO; there were now three Pela Vidda groups (from Rio de Janeiro, São Paulo, and Rio Grande do Sul) and four delegations of the Nomes project (the national coordination project and delegations from Florianópolis, Salvador, and Santos); there were ARCA/ISER and ABIA from Rio, and the smaller ALIA from Londrina, Paraná; there were also several gay groups (Atobá and Turma OK, from Rio, and the Grupo Gay da Bahia), as well as volunteer and support groups such as those linked with the Hospital Emílio Ribas (SP), Solidariedade (MG), Solidariedade (SP), Esperança, PRAIDS, and MAPA.

The enthusiasm of those who took part in those meetings reached its peak with the reading of the letter of support to the Brazilian Network, which was signed personally by the director of WHO's GPA, Jonathan Mann.[66] Several important documents were presented and discussed—the proposal for a chart of principles for the network;[67] the declaration of rights for persons living with AIDS, developed by the Pela VIDDA group;[68] and a document with the statement of principles for the network, which was approved.[69]

The third series of meetings, in April 1990 in Santos, were marked by some "traumatic"[70] moments, which brought insurmountable differences to the surface. A major conflict split the participating organizations and "postponed the dream of the creation of a Brazilian Network of AIDS–NGOs that included everyone, from Oiapoque to Chuí."[71]

Social commentators associate the schism with two contradictory styles of organizing. Some organizations whose constituencies preexisted the AIDS crisis (including the gay and prostitute groups) had been repressed under the military regime and were now willing to work with policy makers toward the prevention of AIDS.[72] Other AIDS–NGOs integrated anti-dictatorship that refused to compromise with a government that seemed to keep the structures of the former regime which they had fought and did not trust.[73] Several other issues were at stake, such as class, regional tensions, and the permanent underlying differences in race and gender. No matter how critical they might have become, these issues were not what counted the most at that moment.

An analysis of the worldwide processes involved suggests that the critical reason for the breakdown of the movement was not so much its internal characteristics but rather the international background to it. In spite of its inconsistency, looseness, and contradictory nature, the movement might have continued if there had been enough funding, strong leadership, and support. However, many local activists would not accept the fact that the period of revolutionary transformations was basically over. The attempt to create organized international networks, such as ICASO and the Latin American Network of AIDS–NGOs, failed because, according to Brazilian analysts, conflicts regarding the representation of groups and countries, allocation of resources, power competition, and bureaucratic obstacles remained unresolved.[74]

The problems were of a wider scope, however. Funding and support were lacking everywhere—for example, in New York City, CRIs were about to lose their funding. With the change in leadership of the GPA, the world seemed to be shifting in another direction. It might have been that the "utopic moment" of transformative action was over, having been replaced by more institutional, bureaucratic, and "medicalized" approaches.

Unfortunately, the epidemic persisted, providing growing evidence that technical responses were not sufficient to stop its spread. And the more that social intervention was needed, the less it was supported. The expectations for the Sixth International Conference on AIDS (San Francisco) in 1990 had been thwarted. NGOs boycotted the conference because of the U.S. restrictions on the entry of HIV-positive travelers, which prevented people with AIDS (PWAs) from participating.[75] It contradicted the entire rhetoric of WHO and the AIDS action movement. The boycott of the San Francisco conference was respected and followed closely by Brazilian NGOs.[76]

The second international meeting of AIDS–NGOs, held in Paris in November 1990,[77] did not increase the strength of the movement either. Instead, for many activists those meetings marked the end of a dream of equality and of their ability to institute change. According to a delegate from São Paulo, the differences in Montreal that had been seen as the impetus for creating more exchanges had been transformed into insurmountable barriers.[78] One of the things that shocked Brazilian activists the most was the exclusionary politics that had been adopted in the fight against AIDS. Two of the delegates from Brazil, who had chosen not to be tested for HIV, were not allowed to attend a meeting that was exclusively for HIV-positive people. At the time, Brazilian groups fought hard for inclusive politics (regardless of one's serostatus), using the line "We all live with AIDS," a motto of the Pela VIDDA group.

* * *

The year and a half between the NGO meetings in Montreal (June 1989) and in Paris (November 1990) were the peak of exacerbation for this social movement. There was a shared belief that something completely new was going on. Jonathan Mann referred to it as a Health Revolution: for the first time, affected communities were heard, fought for their rights, and challenged their own governments about health policies. People with AIDS had taken into their own hands the responsibilities of organizing, demanding, showing directions for research, and assessing their own treatment needs. They had turned upside down some of the old assumptions of medical practice, wherein patients were voiceless bodies in which symptoms were identified and acted upon solely by medical doctors. Now the "patients" were (impatient) persons with a voice, awareness, knowledge, and the ability to act.

The changes brought about by AIDS could have transformed the entire field of health throughout the world; there was now an officially endorsed and globally disseminated awareness that in order to address some medical problems, one had to address questions of poverty, social development, and human rights. The flame of AIDS-inspired enthusiasm faded quickly, however. As in so many other social upheavals, the goal of equity was undermined by the inequities of real life, and the differences within the social movement crystallized into antagonisms that made it impossible for organizations to cooperate with each other. International funding agencies ceased to promote vaguely defined social goals that could be interpreted locally at will, and instead directed quite closely the projects that were chosen for funding. After 1991, the belief in radical transformation gave way to more limited ambitions, short-term goals, and more pragmatic as well as more easily funded proposals.

Internationally and locally, AIDS had became more "domesticated" for the medical establishment; more drugs and therapeutic approaches were available, there was less fear and panic, and health care professionals and the pharmaceutical industry had more experience with the disease. The energy of radical activists had been drained by losses, burnout, and diminishing international funding. Pragmatic alliances—with the government, the medical sector, scientists, or local corporations—appeared as a possibility, however. After 1992, a new National AIDS Program in Brazil was developed through a number of social interfaces and a significant amount of funding from the World Bank, and it helped redefine much of the landscape for local activism.

A strategy that defined a successful platform for expanded cooperation started in 1991 with the People Living With AIDS Meetings organized by the Pela VIDDA group from Rio de Janeiro. Pela VIDDA had been created one year earlier on the impulse of Herbert Daniel, a Brazilian PWA and writer who had perhaps the strongest single influence on the local response to AIDS. Daniel had been a radical revolutionary who had spent several periods of exile in Europe. On his return to Brazil, at the time of Abertura, he integrated a political campaign as a gay candidate for a small ecology party. A brilliant speaker and inspired writer, Daniel could make an audience cheer as easily as he could persuade a funding agency to provide support. His influence touched mainstream international agencies; the Global Coalition's *AIDS in the World*[79] is dedicated to his memory. His fight against the inevitability of death by declaring a commitment to life and solidarity as a person with AIDS (PWA) made him a role model for younger PWAs and brought much energy into the Brazilian NGOs, as well as to international AIDS settings.

One of Daniel's biggest legacies was the Pela VIDDA group, created in 1990 while he was a staff member of ABIA. Pela VIDDA means "for life" and is an acronym for "the Valorization, Integrity, and Dignity of AIDS Patients."[80] The group's headquarters were literally in ABIA's backyard. At that time, ABIA's headquarters were in a mansion in Jardim Botânico, and there was enough space to accommodate the "offspring" organization, with which a symbiotic connection was maintained for years.

The PWA meeting format responded to pragmatic needs and helped participants organize additional actions and projects. A number of factors may have contributed to the success of the new effort: organizations had matured and benefited from earlier experiences; the epidemic had expanded and gained visibility; other social forces were more prepared to interact with NGOs; and people with AIDS, rather than remaining self-assigned experts, were taking on the responsibility of defining the priorities and directions for action.

The formula of the Pela VIDDA meetings remained consistent throughout the following years, allowing them to create a national space for discussing the social issues of the epidemic in Brazil. It also provided a platform for a collective debate on the most important questions: how to live with AIDS, how to implement education and prevention programs, how to negotiate health policies with federal and local governments, how to handle loans from the World Bank and WHO, and how to run local NIH-sponsored projects for the development of an HIV vaccine.

In August 1991, GAPA São Paulo coordinated a successful inter-NGO meeting, bringing together more than seventy organizations.[81] Rather than repeat the failed attempts to institutionalize a consensus, the purpose of this meeting was limited to the discussion of specific issues, such as the vaccine trials.

This type of forum attracted more interest and achieved more success than the previous ones. From then on, Pela VIDDA–RJ organized the annual National Meetings of People Living With AIDS, which brought together delegates from all over the country, along with scholars (microbiologists, clinicians, social scientists, and epidemiologists) involved with AIDS research, as well as international activists, government officials, WHO officers, and delegates from the Global AIDS Policies Coalition.

* * *

As a whole, the social response to AIDS in Brazil was largely articulated by the international movement; the synchrony between global sponsorship and the expansion/contraction of local activism is documented above. Such synchrony was not felt evenly in Brazil, however. Its reverse side was internal dissention. Rather than equalizing local efforts, external aid induced a process of differentiation among local social forces. Some NGOs were more cosmopolitan than others and maintained strong international links; others either played a subsidiary role or struggled in isolation. In the field of AIDS in Brazil, as the international flows were fast, intense, vital in access to information, and involving large amounts of funding, the differentiation process among locals was enhanced. Cosmopolitan NGOs benefited from the international activist energy, expertise, educational materials, vocabulary, graphic styles, and funding. They were able to improve their own networks, and expand their knowledge, visibility, and ability to develop projects and remain active. With greater funding, they were able to recruit more qualified and more highly paid staff and maintain their economic status through grant writing and international networking.

Some local organizations excelled at being cosmopolitan: ABIA, from Rio, maintained a constant exchange with major international NGOs, the GPA, USAID's AIDSCOM (which later merged with AIDSTECH to form AIDSCAP), and major American foundations such as the Ford Foundation and the MacArthur Foundation. ABIA stood out in several international settings: it was central in ICASO and in the Latin American Network of AIDS–NGOs; it had formed a partnership with AHRTAG (Appropriate Health Resources and Technologies Action Group) in the publication of *Ação Anti-AIDS,* or *AIDS Action;* and it was represented at a higher level in the Global AIDS Policies Coalition. Frequently accused of elitism by other Brazilian NGOs (dramatically so during the 1991 meetings in Santos), ABIA would not adopt a populist attitude or lower its standards. Instead, this organization kept giving priority to research and reflected international standards. It combined research with the task of sharing its expert knowledge, mostly through the publication of a nationwide newsletter, *Boletim ABIA* (which documents most of the AIDS social movement), plus a number of target group publications, videos, media appearances, and conferences.

ABIA's unique style resulted from a rare combination of circumstances. With its first major funding from the Ford Foundation, it launched an early international program in 1986. That funding supported an academic style of research on the "Social Impact of AIDS in Brazil," which brought together anthropologists Carmen Dora Guimarães and Jane Galvão and the activist/writer Herbert Daniel. Their work produced a critical examination of the narrowness and cultural inadequacy of the epidemiological models used by the government and local health authorities to report AIDS.[82]

The explicit purpose of the new agency was to "develop a multidisciplinary approach to the issues raised by AIDS"[83] and to attempt "to set forth adequate prevention, education, and information policies."[84] It was part of ABIA's early doctrine that AIDS was a major health challenge with a severe social impact and in need of a multidisciplinary response. It became apparent from ABIA's statement that within the context of AIDS, epidemiological knowledge could not exist separately from social action and intervention:

> Only global prevention action will be able to check the progress of this virus against which no cure or vaccine is yet available. Such an action, however, will only be effective if it manages to avoid the harm caused by fear and prejudice for which only solidarity can provide treatment methods.
>
> The AIDS virus knows no boundaries, nor country or class or social group. The ways by which it is transmitted, through sex and blood, involve complex social practices often surrounded by taboos. This turns the epidemic into both a severe medical and epidemiological threat and a huge historical challenge, raising cultural, political, economic, ethical and legal issues.
>
> There are many facets of social life in every country which are brought to bear on the epidemiological profile of the disease. To know precisely what are the multiple aspects is the fundamental topic of prevention activities. (ABIA 1988a)

For that purpose, ABIA gathered members from all areas of knowledge—medical and social scientists, media and education professionals, lawyers, politicians, artists, and church and community representatives. The stated goals of the NGO were "to follow and assess AIDS-related government policies and initiatives in Brazil." As was made clear, because of the historic neglect of public health in Brazil, "only through permanent surveillance and joint pressure" would society be able "to claim its right to health, forcing the state to take on responsibility in this area." A second set of goals—"to produce and disseminate updated, accurate and reliable information on HIV-infection prevention and control"—was to be achieved mostly through work on information and knowledge: a critique of available information, the avoidance of "misinformation," and the fight against partial and incomplete information, especially information that might produce panic and prejudice; the creation of a means of acquiring direct access to updated data; the preparation of "written and audiovisual aids holding clear and reliable information" and directed at various audiences by ABIA with other agencies; research on the local social impact of the epidemic in order to "establish the epidemiological, social and cultural profile of AIDS" in Brazil; and the provision of consultation and advisory services to the media and in response to public demand.[85]

Combining the executive skills of Silvia Ramos and Walter Almeida, with the nationally known PWA and sociologist Herbert de Souza

(Betinho) as president, and working in collaboration with its many prestigious founders, volunteer associates, and full-time coordinators, ABIA's staff developed a number of international connections that enabled the organization to grow and take on a leading role in the international networks of AIDS. What seemed to the smaller NGOs to be lavish expenditures for airplane travel, computers and modern communications equipment, full-time social researchers, and a number of glossy color publications[86] was routine for ABIA, a fast-acting agency oriented toward working with information, knowledge, and a critical monitoring of the epidemic.

This orientation characterized ABIA's first period, which lasted roughly from 1987 to 1991. During this time, its focus was on broad intervention in the sector of information, including the production of original texts on analysis and reflection, mostly by political writers Herbert Daniel and Silvia Ramos, and international networking, mostly conducted by Walter Almeida, M.D.

Over time, internal changes in ABIA led to changes in its orientation. A second period in the life of the association—1992 to 1994—was characterized by the formation of multiple partnerships with and intervention in different social settings, including workplaces,[87] low-income communities,[88] and schools.[89] Betinho remained as president, while the former staff was lost to either AIDS, burnout, or incompatibilities. The organization's new direction included the appointment of Jane Galvão from ARCA/ISER, João Guerra from NGO community intervention, and Richard Parker from the State University of Rio de Janeiro to the position of main coordinator. Parker, a skilled international fund-raiser, brought to the agency a wave of new funds which reached the level of $1 million a year by 1994. The staff was also expanded to include Veriano Terto, Nelson Solano, José Stalin, Christina Valinotto, Simone Monteiro, Jacques Schwarzstein, Cristina Castelo Branco, Lys Portella, and others. In 1992, the headquarters moved from the elitist neighborhood of Jardim Botânico to a downtown office building. This act was both symbolic and budget-minded, since bankruptcy was imminent at that time. The Pela VIDDA moved with ABIA to the new and more accessible downtown location, which had enough space for ABIA's open resource center, with its documents, videos, and books.

During a third phase, which began in 1994, ABIA's profile changed again. Although it was still in its downtown headquarters, its staff and the number of projects it was involved with were dramatically reduced. Once again, the agency focused on a more academic style of research, as well as on the organization of seminars and the production of knowledge about the social aspects of AIDS. Its projects were carried out in partnership with the university, particularly with the Social Medicine Institute at the State University of Rio de Janeiro.[90] One particular project that combined research and intervention became central to ABIA's activities: the project *homossexualidades,* which targeted gay men, was executed in partnership with Pela VIDDA–Rio and sponsored by AIDSCAP/USAID, the MacArthur Foundation, and the Ministry of Health.[91] After many years of conceptual elaboration and public education against prejudice based on the perceived link between AIDS and homosexuality, ABIA was now ready—with the support of international funding agencies and the government itself—to address the population that continued to stand out statistically in terms of illness and risk of illness.

Only a few other NGOs succeeded in establishing international networks and developing their local action accordingly—that is, on the basis of funding or on the acquisition of privileged knowledge. A successful example is the joint publication by Pela VIDDA–SP and GAPA–Bahia of the *Cadernos Pela VIDDA,* a newsletter on treatment news from around the world which contained a glossary of AIDS-related terms from translated articles. This publication served as a means of sharing a type of knowledge that previously had been available only to a few English readers with access to international newsletters or conferences.[92]

* * *

Consistent opposition to governmental policies regarding AIDS (or, according to many activists, the lack of them) was a central element during the first period of social organizing against AIDS in Brazil. Earlier in the epidemic, public officials had referred to AIDS as a foreign issue with little local relevance, especially in comparison with local endemic plagues. Public dismissal and the government's unwillingness to react to AIDS, until at least 1986, fueled NGO criticisms; this partially explains why their early efforts were concentrated on persuading the public that AIDS was locally relevant and that adequate measures should be implemented. The NGOs assumed that the government would never meet their requests or the needs of the people.

The governmental program on AIDS was created in 1985 and took effect in 1986. During the following years, the program, headed by Lair Rodrigues,[93] collected epidemiological data and published a few manuals for AIDS care that were sent to health services;[94] it also lobbied for international support and scientific attention to AIDS. Working through a loosely organized action program and within the decaying health system that remained in the aftermath of a long dictatorship, the government's action toward AIDS had serious gaps. The government might have had good intentions when it developed its public awareness campaigns, but its efforts were strongly criticized by the most vocal activists,[95] who disapproved of the government's choice of communication styles, its choice of words, its priorities, and the slowness with which it reacted (or failed to react) to critical issues such as blood contamination.[96] In spite of the shared goal of intervening and containing the AIDS epidemic in Brazil, governmental and nongovernmental sectors did not interact well. While the government had a narrowed medical perception of the epidemic, NGOs, backed by international organizations and by their own political and social science knowledge, insisted on addressing the social dimensions of AIDS and on their being the target for a central strategy for action.

NGOs developed their own knowledge of AIDS in Brazil, often rejecting the knowledge used and disseminated by the government, even what we might think of as "technical knowledge." Figures were challenged as being under-representative;[97] initially assuring that the government figures underestimated the number of infections by 30 percent, ABIA later raised the underestimate to 50 percent, assuming that the government's figures for AIDS were even worse for the state of Rio de Janeiro, where they believed that the 800 cases reported in 1989 might correspond to more than 2,000 actual cases.[98] Epidemiological concepts, including the concept of "risk groups," were challenged as culturally inadequate, discriminatory, prejudicial, and misleading.[99] In this area alone, a vast subspecialty of research on local sexual cultures developed, drawing from the constructivist view of local sexual customs,[100] and was developed into an international model by anthropologist Richard Parker,[101] who inspired rejection of the narrower medical models upon which the epidemiology of AIDS was originally based. Government media campaigns were rejected for being offensive, ineffective, or scarce, or for using incorrect imagery;[102] government publishing efforts were criticized for producing too much or too little. Eventually, harsh criticism of the government gave way to a more "objective" analysis of governmental efforts and the possibility of working cooperatively toward containing the epidemic. Until then, the knowledge and world views of activists and public officers seemed to belong in different galaxies.

There were historical reasons for the opposition between nongovernmental and governmental agencies. Since 1964, the governments of Brazil had been military dictatorships, and NGO personnel had matured politically while fighting the government. To the average NGO activist, "the state was an obstacle, rather than an actor for cooperation."[103] Even after civilian rule took effect in 1985, there was no evidence of a substantial change in the method of administration.

The 1990 presidential elections brought a promise of change through labor movement candidate Lula and the anti-corruption candidate Collor. Collor won and proceeded with frantic reforms to the monetary system and multiple efforts to modernize Brazil's international image. He did not, however, change much about health care, except to allow the slow and insidious dismantling of the public health care system, with a diminished role for the state. Ironically, this president was impeached two and a half years after his election, on the grounds of corruption. His first appointed minister of health, Alceni Guerra, had been dismissed earlier for proven corruption—allocating funds for the 1991 anti-cholera campaign for his own electoral benefit, and overpaying for "emergency" purchases, often defined by the interests and profit of his political supporters.

Under Guerra (1990 to 1991), the little that the government had previously done to fight AIDS was replaced by fewer and constantly postponed actions. AIDS officer Eduardo Côrtes, who at the time of his appointment (1990) was more highly praised than his predecessor, Lair Rodrigues, promised fast action and an "aggressive" media campaign. The campaign came to light a few months later, funded by a consortium of corporations. The poster's tagline—*Se você não se cuidar, a AIDS vai te pegar* ("If you don't watch out, AIDS is gonna get you")—combined with naked silhouettes of a woman and a man with targets on their genitals, was considered to be in bad taste and ineffective. Yet it was bland compared to the TV spots of the same campaign. In those, a succession of people with illnesses such as cancer and tuberculosis announced to the cameras that they had been ill but had been cured. At the end, the screen brought in the face of a young man who announced that he had AIDS, and that there was no cure. The campaign was declared a disaster by the NGOs, which were backed by letters and comments from the public[104] and by many of the health care professionals I interviewed. The relationship between governmental and nongovernmental organizations continued to sour. The years 1990 to 1991 were among the economically worst times ever for Brazil; it was during that period that the AIDS crisis became more evident and demanded more action. By 1991, opposition between NGOs and the government had reached its peak.

It took time, changes in the Ministry of Health, international pressure, mounting evidence of the gravity of the AIDS epidemic in Brazil, and the development of an apparent medical and social emergency for some convergence between the governmental and nongovernmental sector to occur. Toward the end of 1992, the two sectors interacted more often.[105] The government created a means of interfacing and negotiating with NGOs and regularly called in consultants and experts from the world of NGOs. New governmental concepts and strategies showed evidence of NGO influence, especially in their communication style, publication style, and interest in community-based organizations. Also, NGOs agreed to negotiate with the government about specific issues, some of them as critical as the support for vaccine development projects or the allocation of World Bank special funds for AIDS prevention and care in Brazil.

Even though their convergence was not symbiotic, the boundaries between them occasionally became blurred. Many activists complained about confused identities when they started receiving funds from the World Bank via the government, saw their organizations listed in government-published catalogues,[106] or heard their own words in government-sponsored campaigns.[107] As some organizations became locked into a survival strategy that brought them under governmental supervision, others moved into other sectors of intervention, such as applied or academic social research on AIDS-related subjects—e.g., sexuality, drug use, social meanings, and coping strategies. International commentators observed that during the Tenth International Conference on AIDS in Yokohama in 1994,[108] the booth for the government of Brazil looked like that of an NGO, whereas ABIA's resembled the booth of an academic publishing house.[109]

At that time, the government had taken over much of the role of defining the world of NGOs—a role that is not unrelated to the general movement toward ridding the state of its social functions by relying on their fulfillment by civil organizations, whether or not those organizations were funded for that purpose. In 1993, the Brazilian government matched a $125 million World Bank loan to fight AIDS using money from its own treasury. The total amount of the program was nearly $300 million.[110] Furthermore, the government passed funds on to different NGOs through cooperative projects. Many NGOs felt coopted and tied up, limited to work on the one project for which they had written the grant proposal; this was a radical shift from the way they had functioned in earlier days, when international funds were allocated for loosely defined social goals. As was the case in many other Latin American countries, the AIDS–NGOs in Brazil were on a "journey from protest to proposal"[111] and did not feel completely at ease in their new social persona.

Calling on the Pela VIDDA and ABIA coordinator Stalin Pedrosa to head the interface with NGOs, the government took on the role of overseeing the world of AIDS–NGOs. In 1994, they counted 140 NGOs serving AIDS, with 69 of them specifically defined around AIDS and/or human rights, 26 associated with a religious denomination, 14 representing women's organizations, 9 serving a gay population, 4 being primarily social movements, and the remaining 4 serving health care professionals.[112] In 1993, 55 of those NGOs had joint projects with the government, having received funds ranging from $6,000 to $100,000. In 1994, 24 NGOs were funded for new projects, and 34 were waiting for approval.

* * *

In Chapter 2, it was stated that one of the strongest components of the AIDS social movement in the United States was its ability to interact with the medical establishment and to intervene in the previously inaccessible world of biomedical research. In Brazil, however, the social movement that responded to AIDS did not include much of that component. Before 1991, when *Cadernos Pela VIDDA* was first published and public discussions on local participation in HIV vaccine development were initiated, there were no traces of local "treatment activism." The changes this disease had brought into the doctor–patient arena were restricted to a niche of educated, personally empowered middle-class patients with access to information. This was where much of the new social character grew: the person with AIDS, fighting for his or her life and anxiously following the latest scientific news to try to transform a "death sentence" into a chronic condition; this had a special impact on a medical culture such as the specialty of infectious disease, whose practitioners were known for their aloofness and distance from the patient (see Chapter 5).[113]

The intimacy with which North American activists worked through partnership-based innovations in science (see Chapter 2) was far removed from Brazilian reality. Any relationship between a Brazilian PWA and a medical drug was mediated by a number of institutions, including laboratories, corporations, agencies, research centers, and information centers that were located somewhere else, and above all, physically and politically, in the First World. People involved with AIDS in Brazil were by definition in a peripheral setting: they were in the Third World, removed from the centers of decision making and doubly removed from the possibility of lobbying the decision makers directly. The ability to interfere in the core of the medical and scientific process was submerged in an ocean of basic and urgent needs. Drugs that were approved long ago were still out of reach because of their expense or their lack of availability in Brazil, and there were not enough hospitals and hospital beds. Furthermore, many doctors and health care professionals still refused to treat people with AIDS, families rejected relatives with AIDS, and an intense fear and prejudice dominated Brazilian society as a whole.

Given the context within which Brazilian NGOs had to negotiate, these organizations focused on more general issues. Like the agencies that supported them, Brazilian NGOs were concerned with the potential annihilation of human rights brought about by the popular perception of AIDS that automatically placed people with HIV infections on death row and stripped them of their rights. Community organizations fought for solidarity, which was defined as the "social vaccine against AIDS."[114]

For a number of years, no actual bridges or compatible knowledge between NGOs and the medical establishment were consistently pursued: NGOs worked in the area of human rights and general social issues, while the medical establishment produced and used its own knowledge. The "blood fight," which epitomizes an early interaction between NGOs and the government around a medical question, was conducted under such assumptions: possessing a basic knowledge of medicine, activists demanded action from the government to satisfy their right to have an uninfected blood supply.

As for other topics, there was no consensus among NGOs. Some considered it irrelevant to demand medical "quick-fix" solutions in a country with such a devastated health care system; they preferred to work within the social and political spheres, demanding that the government restructure the health care system in order to provide stronger and more effective intervention. Others preferred to fill in existing gaps—for example, by helping people with AIDS find medication when the public service bureaucracy delayed treatment.[115] Until the free distribution of AZT launched a public discussion in 1991 and subsequently led to the publication of *Cadernos Pela VIDDA,*[116] there was not much room for questioning and interacting with the medical establishment.

* * *

A turning point in the interaction between activists and the medical world coincided with the discussion of vaccine development, which began in 1991.[117] The need for a community partnership for HIV vaccine development brought about the need for a new style of action for both the medical establishment and the activists. Unprepared for social negotiation of their protocols, medical researchers now had to face community representatives or NGOs, and NGOs had to learn the scientific details of the protocols they were involved in negotiating, which led to further mechanisms of differentiation. Through this process, some community leaders became experts on this issue, and thus were invested with additional symbolic power.

The vaccine discussions brought out many of the difficulties facing those concerned with social aspects of AIDS. Tension developed between the government and NGOs, among the NGOs, between educational prevention and vaccine prevention programs, between those proposing different spending quotas, between the First and Third Worlds, and between private industry and public concerns.

The first public proposal to involve Brazil in global efforts to develop an HIV vaccine was made by WHO in 1991. With both an adequate epidemiological profile and a local scientific community, Brazil seemed like an appropriate contributor to the search for an AIDS vaccine. This suggestion was made at a not quite appropriate time. The minister of health and his AIDS office appointee, Eduardo Côrtes, were under severe criticism by the NGOs: the government's AIDS budget had been reduced, the national program offices in Brasília were empty or unproductive, and the government's "aggressive" educational campaign was considered faulty.[118]

Also, a history of abusive medical experiments in the Third World, combined with a sophisticated anti-imperialist rhetoric, inflamed anger about the possibility of using Brazilian bodies as "guinea pigs" for international experiments.[119] There was no local history of treatment activism or experience with volunteer participation in clinical trials. While U.S. activists fought for the chance to be included in clinical drug trials, which often meant taking dangerous, toxic substances, Brazilian activists rejected the entire concept and distrusted the system.

Moreover, the possibility of preventing AIDS through education rather than through vaccines was still thought to be a viable solution. The idea that "the vaccine already existed, and it was called solidarity"[120] had been broadcast throughout the activist networks. To resort to a vaccine was somehow thought to be equivalent to acknowledging defeat and giving in to "medicalization."

Some of the points raised by vaccine development, such as the basic requirements of unbiased drug trials and the production of community documents, piqued the interest of the community,[121] generating the comment that Brazilian "treatment activism" took its actual first steps in connection with the question of vaccines,[122] following its incipient start around AZT.

The vaccine issue got the full attention of the government in 1992, after the new health minister, Adib Jatene, redesigned the priorities for public health to give AIDS full attention. Lair Rodrigues was reappointed to the AIDS–STD office, which now had a multidisciplinary staff with the ability to interface with NGOs and anticipated an international mega-loan from the World Bank. Participation in the vaccine trials was given a high priority and became the subject of a number of documents and seminars.[123] This could have been the moment of a major transformation, but, at the local and international level, history proved that it was not (yet?) to be.

The integration of different specialties—clinical medicine, virology, immunology, epidemiology, statistics, behavioral sciences—with the diversity of community representatives was a difficult (and still ongoing) component of the implementation of the steps necessary for the development of vaccine products. To add to the complexity, there was competition between the different academic institutions in charge of the process, exacerbated by the promise of benefits such as transference of technology, funds, and international connections.[124]

These issues still characterize the state of vaccine development in Brazil, where both WHO-sponsored and NIH-sponsored projects compete for similar populations, which actually overlap in cities such as Rio. In the meantime, vaccine development in the United States ran into difficulties that led to the suspension of Phase III trials (which evaluate drug efficacy in large populations) in 1994.[125] This suspension directly affected the Brazilian situation; when developing countries are encouraged to test products rejected in the developed countries for reasons of potential danger and lack of safety, the specter of the "guinea pig" is naturally raised. The fourth national meeting of People Living With AIDS, held in Rio in 1994, focused mainly on this problem. It included guests from WHO, the Ministry of Health, the Federal University of Rio de Janeiro, the Federal University of Minas Gerais, the University of São Paulo, FIOCRUZ, ACT UP–New York,

ARCA–SIDA, and most of the Brazilian NGOs. No consensus is yet in sight. The most noticeable characteristic of this process thus far has been its openness to public scrutiny.

* * *

To summarize, the social response to AIDS in Brazil should be analyzed in terms of both its sociological (political, religious, and academic) background and its relationship to international responses to AIDS. These two elements, rather than the actual details of the medical and epidemiological problem, help us understand the particular forms and dynamics of AIDS organizations and AIDS activism in Brazil.

As for the actual historical context, the coincidence between the early development of the epidemic and the generalized economic depression and dismantling of institutions at that time made AIDS in Brazil seem "The Worst AIDS on Earth," or was made so, by the government's ineptitude.[126] NGOs demanded the most, and the government responded the least. Inspired by the international movement and supported by international institutions, NGOs aspired to general health reform, if not a health revolution, but the government could provide only scattered and short-term responses. After a period of difficult and heated debates, opponents became closer and better able to negotiate with one another around pragmatic issues, such as the management of the World Bank loan to fight AIDS in Brazil and the possibility of their participation in the international vaccine trials. Many people and ideas were left out of the process: people who lost their lives to AIDS, who left disagreeing with the changing terms of action; ideals that were too broad to be met by the practicalities of human interaction and institutional competition, and which more easily turned into disenchantment. The history of the social response to AIDS in Brazil, as in the world at large, is at once one of achievement, empowerment, learning and teaching, fund raising, research, publishing, media visibility, cosmopolitan links, rushing and anxiety, and one of loss, grief, and sorrow.

The declining charisma, the multiple losses, the bureaucratization, and the routinization of action and research could not find a better arrival than the triple cocktail therapy, announced as a possible cure for AIDS in 1996. At that moment, conflict gave way to cooperation, and efforts were focused for the most part on making the new treatment available to the people of Brazil. Government leaders and activists finally had something on which they could agree.

FIVE

Central Problems in a Peripheral Landscape

Negotiating Knowledge and Treating AIDS in Brazilian Medical Settings

While local nongovernmental organizations (NGOs) and governmental and international agencies were developing ways to raise local awareness to prevent the spread of AIDS in Brazil, local hospitals, clinics, and medical offices faced a different aspect of the epidemic: taking care of the sickest of those who were already infected. While the agencies were busy formulating prevention messages and calls for solidarity and against prejudice, addressing political issues, gathering statistics, and developing guidelines and plans, hospitals and health care personnel worked nonstop to help people with everything from the cutaneous lesions of herpes zoster or Kaposi's sarcoma (KS) to *Pneumocystis* pneumonia, respiratory failure, toxoplasmotic seizures, or meningitis, to those facing the loss of sight due to *Cytomegalovirus* (CMV) infection or suffering from HIV dementia, disseminated thrush, or incurable wasting.

People with AIDS (PWAs) also took their symptoms to local organizations (GAPAs, Pela VIDDA groups, ABIA, etc.), as well as to the World Health Organization (WHO), the PWA Coalition, the GMHC, ACT UP, AIDS Hilfe, and other international AIDS organizations. Being intimate with the nature of these infections and being able to treat them and talk about them became a part of living with AIDS. A strange and new specialized medical knowledge became part of many people's young adulthood in the 1980s and 1990s.

For those of us with no medical training who nonetheless became immersed in the world of AIDS, those diseases were not just commonplace references but were often associated with a name and a face: Augusto went blind as the result of a CMV infection, the same infection that shrank Ricardo's body to its bones; Isabel had cryptoccocal meningitis and collapsed rapidly; Pedro suffered the agony of non-Hodgkin's lymphoma; Paulo was severely impaired with toxoplasmosis; Jaime died of the strange *Mycobacterium avium* intracellular (MAI) infection before anyone could understand what was happening; Luis developed nerve damage from a herpes zoster infection that became so severe that he could no longer put on his shirt; Neide wasted down to 55 pounds; John had tuberculosis and such severe thrush that he could hardly swallow his food. Also, a medicalized routine became part of so many lives: alarms rang to time the intake of antiretrovirals, antifungals, nutrients, and other mandatory pills; prescription bottles filled drawers; new body accessories pierced chests and arms—ports and catheters for the daily infusions of ganciclovir, foscarnet, or intravenous Zovirax; bedrooms and living rooms became little hospitals, where infusion poles and articulated beds were now furniture; and family members and friends became nurses overnight.

Yet it was in the traditional hospital setting that all of these elements converged. Under the same roof, side by side, sometimes in the same infirmary room with privacy vaguely simulated with translucent curtains, the most severely afflicted and those with the most complex cases came together. They shared the space with a medical staff that came and went, with a nursing staff that came and went, with other personnel with whom they had previously had no personal or emotional links. Inpatient and outpatient care were provided, although under severe limitations.

The globalizing experience of AIDS may have brought a common set of problems to central and peripheral hospitals alike. However, the latter showed the signs of broader differentiations and world heterogeneities. In Brazilian hospitals, the limitations that affected patients and health care professionals were not exactly the same as those experienced to the north, in U.S. hospitals and doctors' offices. In the United States, lack of knowledge about the new disease was by and large the driving force of AIDS activism. Not that there were no social problems associated with class, race, gender, or location. There were indeed, and they have been subject to social analysis (see Chapter 1). But knowledge—or the absence of it—was a central point. It angered activists, mobilized scientists, and kept doctors and patients on edge. The social movement demanded government money for AIDS research, fought for the release of drugs in clinical trials, lobbied science agencies for more drugs and more research, pressured laboratories to cooperate, participated in clinical trials, donated money to help speed up the process, and created agencies to raise funds for AIDS. In addition, scientists and scientists-in-training engaged in the rapidly growing field of HIV research. Everybody wanted to know more. People from all backgrounds found opportunities to build research careers by specializing in AIDS. Activists, journalists, and many members of the public became amateur scientists, constantly cruising Medline, talking about theories, negotiating trials, interpreting blood results, and creating personal and community strategies for treatment and prevention. Knowledge was the center around which everything else revolved: everything was a matter of achieving, sharing, and applying knowledge in order to spare lives, prevent new infections, and avoid further agony.

In Brazil, the pre-existing deprivation was so extensive that the AIDS crisis triggered agendas that differed significantly from those in the United States in their terms, approach, and content. Information about AIDS was proportionally less scarce than material resources. International information circulated more easily than medical equipment and drugs. New articles arrived almost immediately on publication and circulated freely, either as photocopies or summarized in oral accounts. Yet, in the landscape of scarcity and a collapsing network of medical care that I observed during the late 1980s and early 1990s in Rio de Janeiro, medicines as basic and inexpensive as Bactrim might just not be available when they were needed, and many patients could not afford to buy them in a drugstore. People with AIDS had to endure the pain of curable or preventable infections such as pneumonia because existing treatments were not available or were unaffordable. In this respect, AIDS became like many other diseases of poverty, presenting its effects in inverse proportion to socioeconomic well-being.

For those who were in charge of assisting AIDS patients in the hospital, the frustrations were doubled and grew with the expansion of available knowledge. Not being able to administer some of the available and proven treatments created additional stress for health care providers. As if it were not enough to manage the difficulties of treating AIDS, they had to see some CMV-infected patients go blind because they had no foscarnet, or others battling pneumonia because they lacked such basic items as aerosol Pentamidine and oral Bactrim. Within this context, knowledge about AIDS was a relatively abundant commodity: there was more information available than what could be applied in clinical treatments.

Information came through different channels and was as stratified as contemporary Brazilian society. Some places within the medical care system were the best "doors" for new information to come through; eventually, these became the sites that distributed information to other local settings. As we have seen with the NGOs, some local sites were more cosmopolitan than others (see Chapter 4) and had faster and more efficient access to the latest information from international sources. They included university hospitals, where medical students learn, residents receive their training, specialists pursue their careers, and most medical research is conducted. In an attempt to understand whether those "doors" swung both ways—that is, whether there was local research, and whether it was incorporated into the international production of AIDS research—I selected a university hospital in Rio de Janeiro as a primary field location.

* * *

Before the hospital is described, we should remember that Brazil is not a typically disenfranchised developing country, but one with a contradictory tension between extra-developed sectors and underdeveloped ones, in a unique First World-Third World combination. In a way, it presents a unique combination and the typical interactions of First and Third World communities. Medical settings mirror this contradiction, in regard to both clinical work and biomedical research. "Pockets of development" exist within landscapes of underdevelopment, in medicine as in other aspects of society.

This contradiction matches the well-defined social stratification that broadly characterizes Brazilian society. Local elites live by the standards of the developed countries, with which they maintain regular contact,[1] while the masses live by the standards of developing countries. Patterns of morbidity and mortality suggest the coexistence of at least two different countries:[2] a smaller one with the health afflictions of development, such as heart disease, cancer, and AIDS, and a larger one with the afflictions of underdevelopment, such as infectious disease, high child mortality, and, again, AIDS. Both worlds converge through AIDS, as they do at a global level (see Chapter 3). In a snappy description of the health situation in Brazil, the Brazilian Interdisciplinary AIDS Association (ABIA) reminds the reader that

> Among us, starvation goes hand in hand with top technology, producing all the evils of "backwardness" and of "progress." The absence of protein and the nervous ulcer are side by side in the national stomach; Chagas disease and sudden heart disease equally affect the Brazilian heart. (ABIA 1988b)

AIDS may have brought these two worlds together, but access to treatment and medical care separated them again. An upper-class gay man from Leblon[3] and a poor housewife from Baixada Fluminense,[4] for example, might have shared the same AIDS-related illnesses; they might even have shared the same infectious disease specialist, who served at a peripheral public hospital in Baixada in the morning and a private office in Ipanema in the afternoon. Yet their experiences with the disease were quite different. If the hospital's pharmacy ran out of medicine, she could not buy it elsewhere; he, on the other hand, could send an order for the medication by fax to a network of air stewards, who would bring it from New York within twenty-four hours.

Their doctor traveled daily between the two worlds. In the morning, she could not give more than a few minutes of her time to each of the many patients who stood in line for hours after long bus rides from their distant homes, their patience buoyed by the fact that it is so difficult to get an appointment in the public hospital (which was free) that no one would dare miss it by being late, and by the fact that there were always so many more people needing assistance than there was time to provide it. In the afternoon, the doctor did not have to endure that type of pressure; nor did her patients. A generous amount of time was allotted for each appointment, and patients were seen in a comfortable office in an Ipanema high-rise. If the patient got there early, he could sit on a sofa, listen to music, enjoy the paintings on the walls, or read one of the glossy magazines in the waiting room. When it was time for his appointment, he and his doctor could take their time discussing the medical aspects of the disease and suitable treatments, and even elaborate on the effect that AIDS was having on each of their lives. With the other problems resolved, they were limited only by their knowledge of AIDS. They only had to wait for more scientific production, and they followed avidly any news about new AIDS-related products and whether they would work. They cooperated with each other in discussing and choosing treatments and exchanging information about the latest ones. This doctor–patient encounter replicates much of what goes on among the upper classes in New York, San Francisco, and Paris. In the "First World environment" of private medical offices or in expensive private hospitals and clinics, patients can get all that money can buy. The staff knows everything that is known elsewhere, their limits based only on the fact that no one knew enough about AIDS to prevent AIDS-related complications, which devastated the lives of rich and poor alike.

Not everyone in the "pockets of development" was referred to private practice, however; nor were all public services marked by the ultimate deprivation. A number of "islands of excellence" developed within the public sector, especially in the area of research. Local scientific work was carried out mainly by federally or state-funded institutions. This was the combined result of a tradition of federal and state investment in public research institutions and support from the military regime in the 1970s, whose ideology was grounded in the belief that growth and development could not be accomplished unless Brazil had competitive institutions of science and technology. According to the sociologist Regina Morel, during that decade the military government adopted a policy of developing local sectors of science and technology as part of its plan to fully modernize the country.[5] Despite a growing foreign debt, the regime implemented a rapid modernization program that included heavy industry, nuclear power plants, trans-Amazonic roads, state-of-the-art telecommunications systems, and research agencies such as CAPES and CNPq, as well as federal universities in several states.[6]

The tuition-free federal universities, which are open to all who pass the vestibular entry exams,[7] are traditionally the most prestigious places to study, teach, and conduct research. State universities are generally next in the hierarchy, and most private schools come last. Federally funded research institutes, such as IOC/FIOCRUZ (see Chapter 6), are also among the most prestigious institutions and have solid international reputations. As mandated by the Constitution, research funds are handled almost exclusively by the federal agencies CAPES and CNPq, or by the state agencies like FAPERJ and FAPESP. Federal university hospitals, therefore, bring together the best research and medical care, because it is there that the most prestigious clinicians and researchers are concentrated, as they can pursue their scientific careers in those institutions.

* * *

Having analyzed the fermenting power of the interactions between activists and scientists in the United States (see Chapter 2) in changing scientific protocols and WHO-sponsored First World–Third World alliances in promoting the development of new ideas for the fight against AIDS (Chapter 3), the well-funded university hospital in cosmopolitan Rio seemed to be the right place to look for answers to my questions:

- Did two-way interactions exist between local research settings and the central agencies of science, such as WHO, the NIH, the Pasteur Institute, and major universities abroad?
- Did local institutions have any innovative research protocols, like the ones developed for community research trials in the United States?

I assumed that some peripheral research settings in Rio had the potential for interaction parallel to that of the treatment-oriented activist groups in New York and San Francisco. In a vague way, global agencies assumed the existence of that potential as well.[8] It remained to be seen whether such interactions, if they existed, actually produced any major changes in scientific disciplines, either in the form of knowledge exchange or in the actual content of the science they produced.

I also wanted to know whether AIDS research brought about an actual transformation in the type of interactions that occurred between local and mainstream researchers analogous to those that resulted from interactions between treatment activists and scientists in the United States. Traditionally, peripheral sites had been voiceless in the production of science. More than one local scientist has referred to the lack of visibility for Brazilian scientists in international settings, which was often considered a sign of prejudice or the system's inability to include their contribution. One Brazilian scientist commented that they were seen "as nothing, absolutely nothing" (Jacques, 40); another exposed her own perception of how she was perceived by U.S. colleagues when she stated that "they were all surprised that I was a real researcher, not just a Third World native" (Marie, 37).

But medical patients had also been traditionally voiceless regarding their health. The feminist movement changed the order of things, and the fight against AIDS brought further transformations, through the self-empowerment of PWAs and the globalization of efforts against AIDS. Would there also be a new kind of "empowered peripheral researcher," akin to the recently empowered new patient, able to contribute his or her own findings to the world of knowledge making? This possibility was stated implicitly in the rhetoric of global agencies such as WHO, which portrayed the world of AIDS knowledge production as one that included contributions from everywhere, as if every local experience would be compiled in global data banks and made available for multidirectional and democratic exchange. Evidence of multilateral activity appeared virtually everywhere—in oral communications and in the posters presented at international forums sponsored by WHO and other agencies. The exploration of the activities behind those results and the social interactions in the local research settings that produced them became a goal of the ethnographic research I conducted in Rio.

My research problem unfolded in a number of directions and fields of inquiry. Clinical strategies constituted the area in which AIDS treatment activism in the United States had produced the most significant transformations. It had sped up the production of medical knowledge by creating community clinical trials and monitoring biomedical research. These activities required the formation of new types of partnerships between the medical establishment and the rest of society. After struggling over treatment options and bureaucracy, U.S. activist groups extended their influence to basic science as well, and adopted another level of challenge in the interaction with the biomedical world with questions about AIDS pathogenesis and the role of HIV.[9]

Interactions between activists and mainstream scientists had different meanings at different periods of time and for different activist groups. Such interactions either involved the negotiation of research priorities and questions or challenged accepted models, such as the "HIV hypothesis," in favor of a multicausal model.

Areas of negotiation over knowledge about AIDS had an apparently different profile in Brazil. One of the most sensitive (and also the most documented) areas of dissent and challenge in Brazil was epidemiology. Discussions and social negotiations led to the correction of some AIDS-related epidemiological definitions, especially as they affected access by specific groups to treatment choices, clinical trials, and support programs. But epidemiology was not a priority for discussion and transformative action in U.S. settings. In Brazil, however, epidemiology became the subject of passionate literature from social organizations and social researchers, who challenged the cultural assumptions behind the epidemiological model for AIDS.[10] While social categories and cultural assumptions about behaviors that increased the risk for AIDS were shared by mainstream scientists and activists in the United States, in Brazil they were the object of radical critique and stimulated further research. Carmen Guimarães, Herbert Daniel, Jane Galvão, Silvia Ramos, and Richard Parker (see Chapter 4) noted the discrepancies between the epidemiological model that was used internationally for AIDS (which was adopted by the Brazilian government) and the social, cultural, and thus epidemiological realities for people with AIDS in Brazil. That discrepancy, they argued, might alienate the population, keeping them from following prevention messages, and might therefore have a negative impact on efforts to contain the epidemic. One of the purposes of ABIA was to conduct basic social research on Brazilian sexuality and representations in order to address that discrepancy, which, in their words, the government uncritically supported (see Chapter 4). ABIA's prevention efforts clearly tried to move away from the stereotyped risk groups defined by standard AIDS epidemiology. They argued (and ethnographic observation has confirmed) that the concept of "risk" was quite problematic within the local context. Once word about risk categories was spread by the media, they rapidly became loaded with stigma, as if to be considered "at risk" was derogatory in and of itself. Prevention messages addressed either the general public or specific groups. A certain amount of originality was required to target certain groups. For instance, men who had sex with other men were not addressed as gay or bisexual, because many of them would not identify themselves by those labels. Instead, they merely identified themselves primarily as males who were involved in sexual acts (which happened to be with other males). Other target groups were construction workers, sailors, street youth, women, or youth in general.

The Brazilian government, accused by activists of having adopted the international models too readily and uncritically, had its own preferred topic for international negotiation in the knowledge of AIDS: the case definition. Government officers from the AIDS–STD (sexually transmitted diseases) division in Brasília, the modern capital of Brazil,[11] spoke about the role that Brazil and other South American countries had played in the revisions of the standard definition of an AIDS case that was proposed by the Centers for Disease Control (CDC).[12] The revisions adopted by the Pan American Health Organization (PAHO) included pathologies that were more relevant to Latin America. The "Caracas Definition," which established a number of points and scores for AIDS case definition, resulted from those joint efforts.

* * *

What would be the main area of negotiation within a clinical, teaching, and research setting such as a university hospital? Clinical aspects, as for the activists in the United States? Or, as the latter adopted in a later phase, problems related to basic science? Would the alternative immunology models I had learned from Brazilian scientists[13] provide the basis for alternative exploratory AIDS models and possible therapies? How would the tradition of tropical medicine and experience with infectious disease influence our current understanding of AIDS? Would Brazilian doctors question, like local NGOs, the official epidemiology of AIDS or, like the government, the AIDS case definition?

What I would observe throughout fieldwork was that the "fervor" for achieving knowledge and overcoming difficulties and limits had an institutional counterpart in the promotion of forums that combined a diversity of social actors. This included breaking the traditional boundaries of professional specialties and disciplines, as happened in the International Conferences on AIDS and in many multidisciplinary teams throughout the world. But what I would observe, too, was that those temporary transdisciplinary sites born of the emergency of a crisis in knowledge were not to remain in shape for too long. They were soon to fracture along the lines that separate, and eventually oppose, institutionalized disciplines, professions, and social roles.

I also learned that there was no such thing as a monolithic knowledge of AIDS, that is, a local "product" negotiated as a whole in international settings. The ongoing negotiation of certain aspects of knowledge accounted for a number of heterogeneities, and its results did not have a stable, identifiable shape. One of the characteristics of that ongoing process was its tendency to fracture along the lines of previously existing disciplines, despite the efforts to make bridges between the producers of different kinds of knowledge that could lead to a sort of transdisciplinary utopia. The creation of interdisciplinary teams was somehow its institutional side.

The effort to attain transdisciplinary knowledge was present in the hospital setting I observed the most, which had been conceived with an interdisciplinary format. Efforts to gather together different kinds of health care professionals and cross the disciplinary boundaries in order to respond to AIDS suggested an interest in a new "medical culture," one characterized by permanent information exchange and boundary crossing. However, rather than search the ethereal space of transdisciplinarianism, their effort was focused on established disciplines and medical specialties. Parallel networks, rather than a single matrix, interacted with the so-called global efforts. There were simultaneous and varied scopes of interaction that moved at different paces and acted upon different areas of knowledge, producing the same kinds of schisms that already existed between different social organizations, such as the government, the NGOs, and the health services themselves.

Finally, another overwhelming fact affected the production of knowledge in a unit conceived to provide medical care, in a setting where scarcity and absence of quality services placed excessive demand on the university hospital. Drained by exhausting clinical work, health care professionals often felt that they did not have the time to engage in research, which could have converted their experiences into relevant knowledge. Knowledge production was secondary to clinical work in a university hospital, and the quality of clinical work was subordinate to the overwhelming demand that an impoverished public placed on the hospital. That theme would recur in interviews with the hospital staff, particularly doctors who felt torn between the demands of providing care and the demands of research.

* * *

I chose the special AIDS unit of the University Hospital (HU) of the Federal University of Rio de Janeiro as my main research site. This clinical setting was described by many as "out of this world." Its remarkably good facilities were not representative of the Brazilian health system. To compensate for this bias, I observed a few other outpatient and inpatient sites in Rio and Niterói (which is across the bay from downtown Rio) where people with AIDS and other STDs were treated. These sites included the more deprived University Hospital António Pedro in Niterói, which is part of the Universidade Federal Fluminense (UFF); the old Hospital Gaffrée e Guinle in Tijuca, which is attached to the city university, Uni-Rio; the outpatient Health Center in Rua do Resende in the rundown neighborhood of Lapa, where prostitutes and transvestites hustle on the streets; and the complex of FIOCRUZ, a large campus on Avenida Brasil that houses the Oswaldo Cruz Institute for biomedical research and other research facilities, the National School of Public Health, and the landmark Hospital Evandro Chagas, created in the 1910s as a research hospital to study and treat tropical infections.[14] I also made brief visits to medical care and research centers in São Paulo, Bahia, and Belo Horizonte.

In Niterói, I followed an outpatient service for a low-income population and was temporarily incorporated into a team that provided medical care as part of its medical training.[15] In FIOCRUZ, after attending a course on tropical medicine, participating in public conferences and seminars, and interviewing scientists in different research laboratories, I became involved with a research team as a consultant.[16] Most of the time, however, I observed the ongoing processes as an anthropologist, with an eye on the "imponderabilia of daily life" and the routine of social interaction, and the help of interview scripts.

The AIDS Care Unit of Fundão (the popular name for the Hospital Universitário Clementino Fraga Filho [HU], which is part of the Federal University of Rio de Janeiro [UFRJ]), was known to be exceptional and nonrepresentative of the decaying public health system. "Não existe!" ("It doesn't exist!"), people said to express how rare and unique the service was, and to add a touch of skepticism about the public health service in general. Created between 1986 and 1987[17] with a special endowment and commitment to treat AIDS, the unit was able to provide an outstanding quality of service that stood out in the poor landscape of public health services in Rio. The latter were known mostly for being on the verge of bankruptcy, if not already beyond it,[18] and were notorious for their inability to respond efficiently to the increasing demand on health services due to their lack of resources, their overworked and underpaid staff, their decaying buildings, a shortage of medication, a stressful environment, and unsolvable local social problems.

The University Hospital (HU) looked quite unreal, a somehow extraordinary building in a quite uncommon setting. It is located on Fundão Island, the smaller of a set of islands in Guanabara Bay that lie close to Avenida Brasil in Rio's Zona Norte, between the slum complex Nova Holanda/Favela da Maré[19] and the international airport Galeão. This hospital was part of a set of modern high-rise buildings that had been planned since the 1940s and built in the 1970s under military rule as the campus of the Federal University. The buildings were more than walking distance apart and were set in a desolated landscape, far from the urban effervescence of downtown Rio. The campus layout (as with other federal universities of that period) was designed to serve the authoritarian regime's goal by preventing students from interacting outside class, and therefore excluding potentially leftist politics from academic life. The size and structure of the buildings suggest the megalomaniacal dreams of unlimited growth that shaped the 1940s and peaked under the military regime in the 1960s and 1970s.

Projected on a larger scale than could ever be accomplished, with a total area of 220,000 square meters and 1,800 beds,[20] the University Hospital remained unfinished for years, and was derogatorily called a "white elephant."[21] The Federal University medical school was left for years without a clinical hospital of its own, while others with fewer resources or less prestige used more pragmatic methods to obtain their own limited facilities. The State University UERJ obtained the "rejected" hospital Pedro Ernesto in Tijuca;[22] the city university Uni-Rio used the much older Gaffrée e Guinle hospital, also in Tijuca; and the public health system INAMPS took the new building of the Hospital da Lagoa. Construction of the Hospital of Fundão resumed in the 1970s, but the number of beds was reduced by half.[23] Forced to choose between demolishing the excess space and simply leaving entire wings unused, the board went for the latter solution.[24] At the time of my research, some floors were vacant, and their cement structure was fully exposed. Occasionally, services that did not serve patients directly, such as epidemiology, took over unoccupied areas and "squatted" in improvised "offices" by pulling electrical wires from functioning areas and laying carpets, putting in doors and windows, and bringing in desks, chairs, and file cabinets—rather than waiting for the endless bureaucracy to find space for them or allowing a shortage of resources to delay their official move into more adequate accommodations. In that respect, HU mirrored the juxtaposition of Third World conditions and First World technology that Brazilians so often use to depict their country, and which Rio so well exemplifies with its intertwining of *favelas* (slums) and *asfalto*.[25]

Known as "5F," the special AIDS unit was located on the fifth floor of the hospital, next to other DIP (infectious and parasitic diseases) services. The unit was one of the four National AIDS Reference Centers, the others being the hospitals Emílio Ribas in São Paulo, Gaffrée e Guinle in Rio, and the research center FIOCRUZ, where the focus is on laboratory work. Bringing together a number of exceptional people and materiel,[26] the service provided first-rate treatment to AIDS patients within the public system. In the original conception of the service, research was not the first priority. However, a coordinated effort was made to develop the research component. It included the involvement of other sectors of the university in order to enhance the quality of the laboratory work and the knowledge it produced, as well as to promote the engagement of the institution and its staff in internationally grounded research. Target sectors included microbiology, newborn care, and the audiovisuals that were part of the educational material produced in the hospital for the community.

The pressure for clinical assistance was nonetheless overwhelming. The service was constantly pressured by a demand to which it could not respond. The limited number of beds (originally twelve, expanded to eighteen by borrowing beds from neighboring DIP infirmaries) was never enough for the growing number of patients and their relatives who wanted the rare combination of high-quality medical services and the free amenities provided by the unit. Each patient had a private room and bathroom and was allowed a twenty-four-hour visitor, a rare possibility outside expensive private clinics, and a novelty that became difficult for the public hospital staff to administer. There was also a leisure area with sofas and a color TV, recently painted walls and ceiling, and a pleasing atmosphere that contrasted sharply with the cramped and decaying wards of other hospitals.

The suspension of chronic deprivation created conditions for the growth of patient self-empowerment. The health care staff noticed that the mindset and attitude of these patients toward AIDS and toward themselves contrasted with those of patients in the neighboring DIP wards and indicated that patients in the AIDS-specific unit felt more empowered than those treated in traditional clinical surroundings.

Was that "empowerment" extended to the health professionals? Yes and no. Yes, because there was a sense of breaking ground, creating new solutions, and having special resources for that. And no, because the participation on the global process of making science was elusive and feeble, as the ethnography of the unit showed.

In what ways was this unit unique, and how was new knowledge processed in its daily life? In the regular hospital settings, knowledge was based on information provided by textbooks and medical classes, and elderly doctors could be consulted for advice about difficult cases. In the AIDS-specific ward, knowledge about AIDS was constantly being produced. The unexpected mattered—including the social dimensions of disease. Traditionally trained physicians were aware of the social dimensions of infectious disease—from malaria, whose new variants resulted from deforestation; to leptospirosis, which is promoted by poor sanitation, poor sewage, and increased rat populations; to Chagas, which is associated with poor rural housing and poorly supervised blood banks. These social dimensions were discussed in public health forums and epidemiology classes, but they were not supposed to affect clinical decision making. But in AIDS, the social mattered at all levels, and it entered the clinical sphere. There were too many unknown elements. And there was pressure from above to create transdisciplinary teams.

This multiprofessional AIDS team had been created to better address the multidimensional nature of AIDS. It included full-time infectious disease specialists and residents; nurses and nurses' aides; social workers, psychotherapists, and psychiatrists; and nutritionists. Other specialists from the hospital staff—dentists, ophthalmologists, gastroenterologists, dermatologists, pathologists, and pneumologists—served regularly in the unit.[27] The proximity of and constant interaction between the health professions was not a common feature of the hospital culture at that time. Moreover, the organization of this team challenged the traditional, established hierarchies between the health professionals.

The AIDS ward staff held weekly meetings[28] in addition to regular clinical meetings to resolve any difficulties caused by the constant changes in their lives due to factors related to the new epidemic.

During those meetings, they discussed the social and psychological aspects of each case, collectively trying to understand what they could not find by turning to more experienced professionals or reference books for help. They also examined their own problems as a team in the hard task of interdisciplinary teamwork.

Being a novel experience, the interdisciplinary professional meetings did not always have favorable outcomes. Implicit professional hierarchies and differences in relationships with patients persisted in spite of efforts to follow "democratic" practices during the meetings. Those early attempts at bridging the gaps between professional specialties did give the staff a "taste" of the transdisciplinary utopia that existed during that period of time in the global AIDS struggle (see Chapter 3). They also showed the differences between an idealized process of producing knowledge and the actual realities of clinical life.

By 1991, when I was interviewing health care professionals and observing daily life in the unit (then in its fourth year), AIDS was no longer new; the emergency nature of the disease had become part of a "normal" routine. Over time, health care professionals felt that there was less need for interdisciplinary teams, and professional territories became more distinctly shaped. Individuals from the various professions already had an accumulated experience from serving in the AIDS unit. They constantly used and created original knowledge that differed from the knowledge they had gained through their previous experiences as physicians, nurses, psychotherapists, and social workers. I interviewed these professionals using open-ended questions that focused on their experiences treating patients with AIDS, as well as their ability to integrate their experiences and empirically gained knowledge into the rapidly growing global body of knowledge about AIDS.[29]

Could they convert their experience into communicable knowledge, validate it for universal use, make it science? Had they, on the basis of local experience, formulated any hypotheses, developed research protocols, and produced articles suitable for submission to peer-reviewed journals? In other words, were they the social actors implicit in WHO assumptions on global exchanges, the local counterpart of that apparently global and multilateral venture of knowledge production?

Not quite. There had been a period of exceptional transformative change, but throughout the years, as the novelty of AIDS wore off in connection with the accumulation of experience with the disease, the desire to engage in interdisciplinary exchanges began to fade. Health care professionals retreated to their respective specialties as each developed its own AIDS-based subspecialty, thereby sparing themselves the additional effort of crossing boundaries and translating profession-specific concepts for an interdisciplinary audience. With the retreat to traditional disciplines, the transformative fervor (or the potential for that) gave way to a multitude of networks that intervened, at different paces and with varied scopes, at the so-called global level; each of those transnational networks had its own structure, that varied from communicative interaction to dependency and subalternization, and that fractures through professional and disciplinary lines. Accordingly, I organized my data by professional specialties.

* * *

Every professional involved in AIDS care experienced a change. Much of this transformation was related to the need to absorb large amounts of rapidly changing information and eventually produce new knowledge. In the AIDS unit of Fundão, where the extreme deprivation of most Brazilian hospitals was, in some restricted areas, temporarily suspended, participation in the global production of knowledge might have been expected from the special AIDS team. At some level, this expectation was fulfilled: a significant number of team members participated in some sort of research, and many of them sent their findings to international conferences. Although initially considered subordinate to medical care and diagnosis, research became increasingly important to the staff of this AIDS unit and opened horizons they had not foreseen.

Social workers, for instance, experienced a change in the recognition of their role by others. They had been trained to perform multiple tasks and to be rotated through different health services. Those who were assigned to the AIDS unit, however, found that their work was central to the work of the team. The issues managed by social workers—such as understanding patients' life outside the hospital or establishing steady contact with patients' families and partners—were not traditionally subjects of interest to other health care professionals, but in this case they were discussed by all. This team had more opportunities to examine the social aspects of illness. The fact that biomedical knowledge about AIDS was so lacking may have contributed to the increased interest in its social elements. Moreover, there was pressure from "above"—international agencies included—to account for the social dimensions of AIDS. Knowledge of the social aspects of the disease became medically relevant and central to the management of clinical cases, including making the diagnosis, defining its epidemiology, evaluating home care needs, and promoting adherence to pharmaceutical regimens in the home setting. With the integration of these goals into the unit's ideology, the social workers' knowledge, once seen as a residual concern, became medically relevant and necessary.[30]

The growth of the professional visibility of social workers in the field of AIDS had several repercussions. For one thing, they moved from a pragmatic, quick-fix type of work into more engaged participation as AIDS care providers, thereby improving their status among the hospital staff. For another, they developed an intimacy with research, which had not been on their professional horizon before. The move from a relatively unstructured, all-purpose form of patient care into the fully structured AIDS team was accompanied by specialization, described as a move "from the contact with several different pathologies . . . to living, breathing, sleeping, and waking up thinking about AIDS" (Teresa, 40).[31]

The specialized knowledge of social workers was requested by those on several fronts: the health care team, the patients and families with whom they interacted, the community (either at an educational level or from other hospital teams seeking to exchange information), and the international community to which they sent papers describing their work.

Like many other health care professionals all over the country, social workers stressed the fact that AIDS brought them growing experience but at a painful cost. "Every day you see a person dying," they reminded me. They reported the despair and cumulative stress of closely observing the death of patients whose demographic and social profiles were very similar to their own. They had to discover new ways of negotiating problems and solutions within families, who often behaved in unexpected ways.[32] They had to learn how to deal closely and routinely with death, mourning, and terminal illness, as well as how to handle sexuality in broader and more systematic ways. In the process, the specialization in AIDS brought them international visibility and a sense of belonging to the global community. The dramatic change included their conducting original research and submitting papers to international conferences, which gave them a new kind of recognition.

Like the social workers, psychologists emphasized their learning experiences with AIDS—both with patients and through the production of technical literature that opened new professional paths. Unlike social workers, however, psychologists did not report that their role had shifted toward the center of case management; rather, it had developed a specialized niche. The assessment provided by psychologists and psychiatrists as part of a health care team was needed both for diagnostic purposes and to make certain decisions about patient release or treatment. Their main role, however, was a therapeutic one: they helped patients cope with a disease that brought them rejection, fear, mourning, guilt and blaming, experience of stigma and prejudice, and the need to address their own sexuality or substance use. For many patients, the therapist was the only person they could talk to freely and who could help them come to terms with themselves, their conditions, their sexuality, their bodies, and their families, as well as their anger, guilt, and fear.

Psychotherapists reported their experiences with the first cases of AIDS they assisted in the hospital, before public and social efforts to raise awareness were undertaken, and before community support groups were available for people with AIDS. On top of the problems common to terminally ill cancer patients, which were described in the literature, people with AIDS had the burden of stigma attached to their disease, which enhanced their fear of the unknown and was more likely to cause them to panic. Guilt and fear of disclosure only worsened the already complex therapeutic picture, and depression was a common result. The problems brought on by AIDS were challenging for psychologists and psychiatrists. Much of their early work in this context was exploratory; they worked through it by discussing such problems during team meetings.

Some of the difficulty they experienced with those early cases was related to the novelty of the disease and the extreme panic and prejudice it induced. Infected transfusion recipients wanted to stress their "guilt-free" relationship with the disease, the burden of which they considered doubly unfair. Speaking out against the blood transfusion system, the illegal commercialization of blood, and the lack of supervision, they had some instruments to institutionalize their anger. Men who had been infected through sexual contact would seldom admit it and were more likely to attribute their infection to transmission from female prostitutes than to sexual contact with other men. "By denying the means of transmission, they denied to themselves the fact of being sick," said one psychologist. The same psychologist admitted, however, that over the years, patients gradually learned to better accept their sexuality.

With time and with the accumulation of cases, the "novelty" of AIDS diminished, along with the sense of panic associated with it.

Some aspects of the disease became more predictable—patterns emerged, the health care team grew in experience, and public awareness campaigns began to have some positive effects. Anti-prejudice work was implemented by patient-based NGOs, which spread the word that it was possible to live with AIDS instead of feeling that it was an internal time bomb. The media took a less terroristic approach and gave AIDS a human face (see Chapter 4). As a result, patients began to respond to the diagnosis less dramatically, and psychotherapists were able to work with them on other levels to help them learn how to live with AIDS, either as outpatients or in the hospital. The psychologists eventually became familiar with ways of establishing links between psychological health and changes in immune function based on empirical knowledge shared by activists and through discussions that appeared in academic journals.[33] I asked about the possibility of correlating the two levels. Psychologists had an empirical "feel" that there was a correlation, but they would not turn it into a theory. "This is a damn virus," one of them said, holding HIV responsible for most of the damage to the immune system of people with AIDS. Overemphasizing the power of the will to survive, as some activists did, was also a way of denying the condition of being sick with an infection that had no absolute cure.

Like other AIDS care providers, psychotherapists became burned out by their work. In a reflective, psychoanalytic approach to help health care workers serving people with AIDS, they commented on their own distress. According to one,

> It seems like a cumulative and insidious type of stress. We can handle the difficulties of the clinical setting, give the patients their diagnosis, handle their reaction, help them sort things out and make their choices, come back home and still be human, learn a lot from the entire experience, grow scientifically and participate in the international community, but there is an insidious level of stress that we have not come to terms with yet. (Julio, 32)

Some noticed changes in their own health as a result of their work; excessive sleepiness, for instance, was interpreted as being a result of shifting the fears of their patients into their own subconscious.

Psychotherapists and psychiatrists handled a new type of knowledge, gained empirically through a review of international literature. Like the other professionals, they knew they could not rely on experienced supervisors to answer their questions, because AIDS was too new. They also knew that they could not rely fully on their literature review, because the international literature did not cover the diversity of situations they faced. The psychotherapists on the team were the first in their profession to have direct experience with the local aspects of the disease.

They had different perspectives on how to make their unique experience available to others. Some felt that training interns was the best way of imparting knowledge; others decided to teach; still others actively explored publishing or presentation opportunities. For example, one staff member was eager to share with the international community his findings about the scales used to objectively measure HIV dementia. Those scales, which were developed in the United States, had to be adapted because of differences in literacy (one of the factors used in the existing dementia scale) between the United States and Brazil.

Nurses and nurses' aides managed the one-on-one physical tasks required in AIDS care. Doing what most people outside the hospital would not do—such as touching patients and helping them cope with the physiologic effects of multiple infections and vanishing strength—brought to some nurses a sense of professional accomplishment.[34] This was not easy, however, as they had to endure on a daily basis the stress of having to make difficult decisions for patients.[35] Rarely indulging in public visibility or research,[36] the nurses stuck to their vocational roles as health care providers.[37] Crushed between an outside world of prejudice and morbid curiosity about AIDS and their performance as hospital nurses, they were the most susceptible of all the health care professionals to burnout. As a result, they asked for replacements more frequently than the other members of the team. Nurses' aides, who handled the riskiest activities (including venipuncture and the collection of body fluids) with the least amount of professional reward, felt the most vulnerable and were likely to ask to be transferred to other sectors.

In spite of all the efforts toward interdisciplinarity and teamwork, there was a special place assigned to doctors. In the patients' eyes, doctors were still considered "special," above the rest, and therefore entitled to the last and most revered word. Other health care professionals complained that they sometimes had to involve a physician in order to communicate efficiently with the patient, even when the message was related to nutrition, psychological issues, or other nonclinical issues. Interdisciplinarity was a challenge that was experienced differently by each professional I met on the team. While some felt it was an opportunity for growth, not only for themselves but for the practice of medicine and for the patients' well-being, others behaved as if it were an obstacle delaying their work as physicians.

Who were the doctors who rapidly became specialists in a disease that had not existed when they went through medical school? What type of "medical culture" did they represent? Where did they stand within the symbolic hierarchies that organize the medical system? Generalizations about this medical subspecialty should not be too ambitious; the epidemic and its effects kept and keep constantly changing. The ethnography and micro-history of particular settings, such as the one in question, may, however, contribute to the understanding of general questions.

In the unit I observed, as well as in many other places, most of the "AIDS doctors" had been trained in *infectologia* (literally "infectology," or Infectious Disease specialty), also referred to in Brazil as DIP (for *Doenças Infecciosas e Parasitárias,* Parasitic and Infectious Diseases). The allocation of AIDS to this specialty was a matter of debate and was sometimes challenged by other specialties. For example, dermatologists—who treat sexually transmitted diseases—examined some of the early cases of AIDS and also declared it to fall under their specialty. In some hospitals, it was assigned to the oncology service; after all, one of its earliest manifestations was as Kaposi's sarcoma. Internal medicine claimed that there is no particular reason to treat AIDS within any other specialty.[38] AIDS patients, they concluded, could be treated by generalists in consultation with specialists; using the same logic, they claimed that AIDS patients did not have to be clustered in special units.[39]

AIDS had a strong although paradoxical effect on DIP services. Most DIP doctors considered their specialty to be most suited for people with AIDS, which they saw as an "infection surrounded by infections." Also, their propensity for scientific research (see Chapter 6) seemed appropriate for a disease that had triggered so much research so quickly. Moreover, they claimed that they were the ones who took care of the initial cases when it was not yet clear that the epidemic was here to stay, and that it would be a vehicle for research funds and professional growth. Accustomed to "unattractive" and life-threatening conditions in their patients, DIP doctors took the first AIDS cases at a time when, they made a point of noting, no one else would accept the challenge.

On the other hand, a disease such as AIDS was precisely what they had been trying to avoid when they chose the specialty of infectious disease. They had been lured by the efficient and definitive type of intervention that this specialty provided, and by the feeling of "omnipotence" in being able to "actually cure someone who seemed to be dying": "You see a patient one day trembling in seizures, everybody thinks he is going to die, and you diagnose meningitis, or toxoplasmosis, and you give the correct medicine, and soon he is out of the hospital, as if nothing happened . . . that is very rewarding . . . your knowledge brings omnipotence" (Jorge, 30).

They *could* cure. They could play a role in the game of life or death. They could bring back to health a patient who had been dying of meningitis, leptospirosis, or malaria. In their accounts, the DIP patient was the one who either died fast or was cured. Their specialized training provided them with the ability to make an accurate diagnosis and provide adequate treatment, and therefore a cure. The patient was supposed to leave their care free of symptoms and with no expectation of returning to the hospital—at least not with the same ailment.[40]

They knew that their colleagues in charge of patients with cancer or kidney disease had a different experience, that they often had little to provide their ailing patients. And that was precisely what DIP doctors wanted to avoid: the chronic patient, the person they could not cure and whose slow agony they had to share without being able to stop it.

AIDS disrupted their world and their professional self-perception. Omnipotence gave way to a feeling of impotence, humility, and acceptance of their limitations. The impersonal and nameless patient, identified as the carrier of a known infection, was replaced by a complex, demanding, and needy human being whose return to the hospital was almost guaranteed. The conventional short-term and superficial interaction with patients gave way to longer and often emotional encounters and resulted in a doctor–patient bond, which presented new challenges for both parties. Issues that had never been addressed in medical textbooks and courses—such as sexuality, drug use, and domestic relationships—turned out to be central to case management and for developing a rapport. These doctors were faced with the need to develop an entirely new patient-oriented agenda.

The epidemic touched other aspects of their lives. The new patients had social and demographic profiles similar to the doctors' own, completely unlike those of the semi-literate, low-income DIP patients they had become accustomed to. That fact brought the epidemic closer to home. In a previously unknown way, the problems of the patient reverberated in the physician's identity. "It could be my father, my brother, myself," some pointed out. Age, social background, and being sexually active brought the doctors closer to these patients.[41] The detached, aseptic, hierarchical doctor/patient relationship was replaced by constant interactions and circulating questions. The fact that many of these doctors opened private offices (which they could not do as DIP specialists) was a sign of the shifting character of the culture of that medical specialty within the context of AIDS. Interaction with people with AIDS and AIDS organizations changed the lives and perspectives of a number of doctors as well. Learning about gay lifestyles and being forced to become intimate with healthy people who would experience a significant decline in health while in the prime of life brought major cognitive changes to many doctors. The frequent travel to attend international AIDS conferences and workshops (see Chapter 3), the fast growth of international networks of AIDS research, and the availability of funds for international AIDS work all had an effect in the personal and professional lives of these specialists. In the end, AIDS helped transform a low-profile practice into a highly visible profession.

The group I followed was a special one. In addition to the existing DIP staff, the AIDS unit hired eight physicians who had just finished their medical residencies in infectious disease. They selected that specialty before coming in contact with people with AIDS; they saw their first cases during their residencies or immediately afterward. This was a transitional period in the epidemic, when the impact of AIDS was greatest. The generation that followed them would learn about AIDS while still in medical school. This intermediate group became the core of local specialists on AIDS. They were young, just starting their careers; many of them were working on their master's degrees in tropical medicine or infectious diseases. Some would later engage in international research careers, become involved in public administration, or open a private office—rarely an option for infectious disease doctors, but demanded by those who became specialists on AIDS.

AIDS entered their lives very rapidly. They recalled their first experiences with the disease—some only as far back as 1985, while they were serving in this unit; others earlier in São Paulo, where some of the unit doctors had done their residencies:

> I did my residence in Emílio Ribas, which is the large hospital in São Paulo where all the [people with] infectious diseases go. It is known for people crossing the street to the opposite sidewalk, when they pass by, for fear of catching an infection . . . I saw the first cases of AIDS there . . . we did not really know what it was about. (Cecília, 29)

> In the beginning, we used all the paraphernalia—we used surgical masks, we used gloves, long hospital gaunts, all those things . . . we did not know much about in-hospital transmission. . . . my family teased me saying that I was *grupo de risco*. I used to say I was of *grupo de arrisco*. I took the risks

> . . . then we relaxed with the procedures, there were not always gloves and masks to use, then we got used to leaving them behind, we lost the fears. (Jorge, 30)

Some became acquainted with the disease during their residency in Rio, in this or in other hospitals:

> [The supervisor] called me to go see this case. It was a case of immune breakdown. The patient had all these different infections . . . it was the first case of AIDS I saw. We thought then as if it were a rare disease, not as something that was going to come and stay with us the way it did. (Mariza, 30)

One doctor commented that the very first patient she saw as a resident, back in 1985 or 1986, was a person with AIDS, so this disease became immediately incorporated into her professional experience within that specialty.

Combining a rising number of cases with increasing public awareness and political momentum, the SIDA/AIDS program launched in the hospital in 1987 was mainly responsible for making diagnoses and providing medical care. Because of the very nature of the endeavor, handling new knowledge was one of its daily basis tasks. Unlike other infectious diseases, which were fully characterized in textbooks, studied in medical courses, and supervised by elderly specialists, everything about AIDS was new. Textbooks had limited and insufficient information. The older staff members, with greater clinical experience and common sense, did not know the details of the new disease. The team had to break ground, seek knowledge, and be constantly creative.

Asked about their sources of information on AIDS, the team referred to medical journals and peer interactions as the central strategies for keeping up with the field. Traditional clinical and infectious disease journals were quoted the most, including the *Annals of Internal Medicine, Lancet,* the *New England Journal of Medicine,* the *British Journal of Medicine,* the *Journal of the American Medical Association* (JAMA), the *Journal of Infectious Diseases,* and *Reviews on Infectious Diseases.* Less commonly used references were two publications devoted specifically to AIDS, *AIDS: The Journal of Acquired Immune Deficiencies* and *AIDS Research.* Less often quoted but still used by some were general science journals such as *Nature* and *Science.* The degree of intimacy with the journals varied; some team members had personal subscriptions, some consulted them regularly, and some looked only for specific articles mentioned by colleagues. Between the library at UFRJ and the library of FIOCRUZ, most of these journals were readily accessible. Moreover, there were groups organized to photocopy relevant articles so that new information could be shared rapidly and disseminated efficiently. The usual scarcity of written information that affects researchers in peripheral settings was largely compensated for. Information, at least from articles published in mainstream medical journals, was abundant and generously shared. Alternative information, such as that handled by international PWA groups and local NGOs, was not as readily available. Even though some health care professionals read the NGOs' newsletters or participated in some of their activities, those organizations did not provide a source of regular information that was relevant to the practice of medicine.[42]

A number of my questions dealt with the extent to which the mainstream, American- and British-published knowledge fit the local clinical cases adequately. I had been told by government officers that there were some inadequacies in the criteria used for clinical case definitions. Also, being acquainted with the social and activist critiques of the "imported models," I was interested in knowing if there had been any basic challenges to the prevailing clinical knowledge of AIDS. I also asked what they felt was unique about AIDS in Brazil specifically:

- How often had they faced "unexpected situations," and what did they do when they were in doubt about their approach?
- What might be particular and idiosyncratic about Brazilian AIDS? What should the priorities for research be?
- Had they turned their acquired experiences into articles, conference presentations, lectures, or social interventions?

Most of the physicians who responded acknowledged that they had to face new situations on almost a daily basis, and that a lot of creativity was involved in the process. However, they felt that the knowledge created by their experiences was empirical, fit for oral transmission but not for publication. Asked whether their "local knowledge" could be converted into hypotheses for research, most of those interviewed expressed uneasiness with the world of research. Along with material obstacles—such as a lack of money, time, and infrastructure—cultural and psychological obstacles prevented them from conducting formal research and becoming full actors in the global production of knowledge about AIDS.[43] In medical school, they had not been trained to design and carry out research projects—they did not know the procedures, they were unfamiliar with research design and methodology, and they did not feel comfortable writing in English (by and large the language of medical research). Moreover, a background ideology established a sharp separation between their Third World condition and the "First World style" of mainstream medical journals. Many, however, had more basic problems, such as a lack of time and exhaustion from draining clinical work.

Not all of those interviewed were uneasy with research, however. Some had spent time doing graduate work abroad. A period of scholarship abroad had apparently made them more confident about their ability to contribute to the world of "science making." Having an exceptional local laboratory and a full-time staff, and counting on the flow of AIDS research funds from around the world, some foresaw the feasibility of local research. They acknowledged, however, that they could not assume that their unit was "an isolated island of excellence while outside chaos is installed" (Mariza, 30). A lot of other elements—from low wages to the stress of clinical work to the absence of a reliable infrastructure—hindered their research efforts. There was always that background "noise" of being in the Third World while attempting to do First World work, always under the threat of failure or sabotage.

Partnerships with centers of research in developed countries represented an alternative to the more complex endeavor of conceiving original research. But it also takes some extra effort to persuade central research agencies or universities to form such partnerships with them. One local research coordinator once confessed that after frustrating attempts to engage First World agencies in local research, he had concluded that they were more comfortable patronizingly donating funds than establishing a professional relationship with local researchers. After some negotiation, however, several protocols were developed, which brought in more funds, research energy, and space in publications.

In the end, the production of knowledge depended on a complex negotiating process. As a participant in multi-center studies, the Unit found that it was relatively easy to feed data into the mainstream, even though in a quite different manner than I had hypothesized when envisaging the scenario of a paradigm shift in a global setting; that would imply changing hypotheses and research questions, which means being involved in theorization rather than merely feeding data. Also, developing research that was more feasible in Brazil than abroad was another way of achieving visibility. Several articles written by Unit staff members were accepted by peer-reviewed journals, including the annals of the International Conferences on AIDS, whose *Books of Abstracts* became important reference sources.[44]

* * *

If I had expected an ocean of creativity in pursuing the most varied hypotheses on AIDS etiology and pathogenesis, or local clinical and epidemiological aspects of AIDS, if not groundbreaking research on basic immunology, I would have to come to terms with my own naïveté. This was not the site for it. The participation of these health care professionals was bound to add bits and pieces of knowledge to a large building whose architecture had been defined elsewhere.[45]

Overburdened by clinical work, physicians hardly had the time to develop hypotheses. Having access to an abundant although monolithic knowledge of AIDS, they dared not waste time pursuing speculative alternative models. They did not feel entitled or empowered for that purpose. Unlike New York AIDS activists, who were exposed to diverse informational sources and whose basic attitude was tempered by their distrust of the science agencies, and Brazilian activists, who opposed the government, the local doctors I interacted with were tied to a single source of knowledge: peer-reviewed journals. And they depended on it to do their best at their task of treating people with AIDS.

As an example, I pursued the history of their personal understanding of one of the questions, which is revealing about the processes of debate, dissent, and consensus achievement: Was HIV the single cause for AIDS, or should its etiology be thought of in terms of multiple factors? The question had already been discussed by social commentary.[46] I asked doctors about a co-factorial hypothesis versus the prevailing unifactorial hypothesis, centered on HIV.

The question was irrelevant for a number of them, as their first contact with AIDS was at a time when the unifactorial model had been established. They had not followed the Greenwich Village debates (Chapter 2) on possible causes, and most of them were persuaded by the viral hypothesis. For those who remember reading early articles on AIDS, before LAV and HTLV-III were isolated, the viral hypothesis seemed from the beginning the most plausible one, due to the similarities between AIDS and hepatitis B:

> When I read the first articles, it immediately seemed to me that it was an infectious disease, most likely virus related. Its epidemiology was too similar to hepatitis B. (André, 37)

> I was still a student, and we had this science conference of the medical students, and I remember we debated heavily. Some people thought AIDS was just homophobic propaganda from *Globo,* but it clearly seemed to me that there was a transmissible disease that might affect the entire population. (João, 34)

> I did not risk saying it was a retrovirus, I could not have known. But I was quite clear it had to be an infectious agent, most likely a virus, a new one. . . . When Markito [the first publicized case of AIDS in Brazil] died, there were still people that thought that AIDS affected homosexuals because they took hormones and that type of thing. . . . I always thought it should be a new virus. (Claudio, 45)

However, some raised a few questions about the role of the virus in the full explanation of AIDS:

> There have to be cofactors! How do you explain that some people get sick after one year of transfusion and others remain healthy for many years? (Heitor, 35)

Some of the doctors I interviewed wanted to see more research done on the immunological aspects of AIDS and lamented the fact that immunological research is even more expensive than virological research characterization and studies of seroprevalence.

On one thing most agreed: clinical research and epidemiological surveillance were the most "doable" at a local level. If anything, Brazilian AIDS literature would differ from international literature in showing that some local infections, such as tuberculosis, would coexist in a greater number of patients. Some commentators noted that even the tuberculosis research, which was of genuine local interest, was being carried out in First World terms, with expensive cultures and assays instead of the more accessible differential diagnosis method. Research on more affordable means of diagnosis should therefore be a priority and may serve as a way for Brazil to contribute to the global study of health care.

As for epidemiology, they admitted that local forms of transmission may be somewhat unique, owing in part to local patterns of sexuality. That did not mean that they agreed with the social commentary on Brazilian sexual culture that was being reproduced in academic books and in the popular media at the time.[47] Nor did they accept the concept of the "Africanization" of Brazilian AIDS,[48] which pointed to a "heterosexualization" of the transmission pattern (WHO pattern II) and a "pauperization" of the epidemic.[49]

The original Brazilian epidemiological pattern, which was described as similar to that of the United States (WHO pattern I), remained valid to them, because even though they treated an increasing number of women with AIDS, the majority of cases still occurred in gay men. Physicians challenged, however, the notion of "risk groups." It might have accommodated the initial analysis for epidemiological surveillance, but it did not describe what really mattered—behaviors and acts of risk. In their initial contact with the disease, when risk categories were part of the definition of an AIDS case, they were quite concerned with finding the patient's risk group. Eventually, they stopped considering it as central to case management and often left the identification of risk untended. At the time of my interviews, the AIDS epidemiology division of the board of health was concerned with the fact that the category of "unknown risk" was increasing, probably as an effect of doctors' disregard for the problem. In fact, as far as I could determine, AIDS physicians were split in their feelings concerning their lack of interest about the way the patients were infected and their responsibilities toward public health and accurate reporting.

* * *

Within this diversity of quasi-contributions, only a few dissenting views became known, and even fewer were incorporated into international knowledge. The most important were probably those related to tuberculosis and its relevance to AIDS and revisions in the AIDS case definition. Yet, no major challenges to the mainstream production of knowledge of AIDS came from the clinical settings I followed. The effect of the clinical aspects of the disease on the practitioners was too distressing, the pace of production of new knowledge in the mainstream international journals was just too overwhelming, and the self-confidence of those who served at those sites was too low for them to believe that they could participate in the production of science, except as collaborators. As one doctor once pointed out, "We can only find what we were looking for." Sometimes, after new international findings changed the way of perceiving an issue, they could trace back their own perception of it, which they kept only for personal reflection. Even under those circumstances, ideas for innovative research were around, but they were not suited to be turned into research projects, at least not at that time and place.

The two-way interaction and the global circulation of information, as found in a local setting, was asymmetric. Information abounded, but its flow was rather unidirectional. Information was filtered through stable international channels, but pre-consensus debates could hardly be followed, much less integrated. Information that flowed from centers to peripheries became monolithic and uniform; responses to it could be made only in the very same terms, or they would not be heard. As a result, the flow of information from peripheral settings to the centers was scarce and quite monolithic. The global venture of WHO (see Chapter 3) might not have a counterpart in medical research, or at least not yet. The hypothesized "empowered peripheral researcher" had not yet entered the field.

It was farther away from the stressful and constantly draining field of AIDS, or at least farther away from the clinical care, that I would find the equivalent to the imagined empowered peripheral researcher. It was close enough to AIDS, but not quite within it, in the surrounding fields where knowledge is produced: in research laboratories where immunology and infectious diseases were studied, in epidemiology, and in the social sciences. With that finding we get a sense of how the massive globalizing process may have opened new avenues of dialogue, but not as many as it could. Some of them deserve further attention.

SIX

War Metaphors in Germ Theory and Immunology

In Search of a New Paradigm

In the preceding chapters, I tried to avoid the exhausting repetition of the warfare imagery that pervades the field of AIDS, often turning medical texts into assemblages of military reports and militaristic sound bites. Still, the "report from war" genre would well match the feelings of devastation and loss experienced with AIDS. For in so many ways it seems like a real war, with bodies falling, losses being counted, a nightmarish reality, and the expression of euphoria in reaction to any sign of a possible coming cure. But the sense of war is not just about the feelings of devastation and collective loss experienced throughout the epidemic. The language of warfare and its mindset, symbols, imagery, and attitudes are deeply embedded in the several layers of the universe of AIDS, from the macroscopic level of politics and policy making to the microscopic level of biological interaction between cells and viruses.

Warfare is the central paradigm from which everything else in the field of AIDS derives. Action related to AIDS is referred to as a global fight. "War" does not just stand for the confrontation between humankind and a global epidemic; the military language also pervades clinical and lay references to the disease. AIDS is characterized as the result of the action of an insidious virus (HIV) that invades and attacks the human body. HIV disables the body's "defense mechanism," the immune system, by destroying its core elements, the messenger lymphocytes. With its sentinels defeated, the body is left without protection against the "enemy," which under normal circumstances would be rendered harmless and easy to defeat. The body whose T cells have been destroyed by HIV is now abnormally susceptible to the organisms responsible for pneumonia, toxoplasmosis, rare cancers, animal infections, and tuberculosis. Devastated by hordes of microbes, the body ages, shrinks, wrinkles, agonizes, and shuts down.

AIDS therapies are designed to be chemical counter-weapons in this microbiological war. Most noble of all, the antiretrovirals are designed to annihilate the virus that is the "source of all evil": HIV. Since a virus is not really a living creature but a simple, fully characterized chain of abstract chemical information—whose segments have inanimate names such as ENV, POL, and GAG—the antiretrovirals do not really "kill." Instead, they disable one of the steps required for viral replication. The first generation of antiretrovirals—AZT, ddI, ddC, 3TC (the nucleoside analogues)—targeted the enzyme reverse transcriptase. The second generation of antiretrovirals targeted protease, an enzyme responsible for a different step in retroviral replication. The famous triple cocktail, announced in 1996 as a potential cure, combines agents that inhibit both transcriptase and protease. The logic behind the development of the cocktail expresses the logic of warfare: "Hit HIV early and hard," as the scientist behind the medical discovery, Dr. David Ho, likes to say.

These drugs are still quite rough and do not match contemporary techniques of war. The first generation of antiretrovirals might be more accurately described as hitting blindly around their target; they may be better compared to clumsy medieval cannonballs than to the "elegant surgical weapons" of the Gulf War. AZT and its analogues also shot at their target in the dark, destroying both healthy cells and nonprimary "enemies"—microbes responsible for the secondary (opportunistic) infections associated with AIDS. Their efficacy was a matter of debate. Most clinicians argued that they improved the quality of life for the patient and delayed the onset of AIDS symptoms and opportunistic infections, even if they did not seem to prolong lives. Others argued that they did not fully disarm HIV, reverse the illnesses, or restore the body to health. In the absence of better alternatives, however, most people with AIDS (PWAs) preferred to take AZT, despite the risk.[1] Clumsy and inaccurate though they may be, AZT and its analogues were the prime products of biomedical research using the most technologically advanced techniques. Even poorly endowed medical services, such as those I observed in Brazil (see Chapter 5), preferred to allocate some of their limited resources to the purchase of AZT rather than deprive patients of its limited usefulness. People with AIDS in Brazil went out of their way to raise money for a bottle of the drug,[2] and activists included the demand for its free distribution in their political agendas.[3] Only those whose beliefs were based on radically different models (see Chapter 2) chose not to take it. They claimed that its illusory effects were due to its powerful toxicity, which destroyed the agents of opportunistic infections.

The triple cocktail inspired a larger consensus than the earlier drugs. Also, it came on the scene when most of the social forces involved in the fight against AIDS in Brazil were more concerned about cooperation than confrontation (see Chapter 4). Making up for lost time, the government made an extra effort to make the new drugs available through the public health system.

In addition to anti-HIV drugs, a number of antimicrobial agents have been added to the therapeutic "armamentarium" against AIDS. They include additional antiretrovirals such as foscarnet (which has the proprietary name of Foscavir) and gancyclovir (which has the proprietary name of Cytovene), directed against *Cytomegalovirus* (CMV). They also include acyclovir (with the proprietary name of Zovirax), which helps control herpesvirus infections. There are also antibiotics to fight bacterial infections and antimycotics to fight the many fungi whose growth is facilitated by AIDS. As soon as they came out of the test tube, each of these "weapons" became the focus of a fight for FDA approval, release, and price (see Chapter 2).

As illustrated in Chapter 5, infectious disease regained its prominence in biomedicine because of the AIDS epidemic. Being able to control the multiple infections associated with AIDS became the psychological and professional reward for infectious disease experts who could no longer cure patients, as they had been accustomed to doing with patients with other infections. "Winning smaller battles within a broader war" was a common metaphor used by DIP (infectious and parasitic diseases) doctors treating people with AIDS (see Chapter 5).

* * *

The warfare imagery in medicine[4] predated AIDS by many decades, going back as far as the days of Pasteur—that is, to the days when germ theory was being developed. The idea that germs were microscopic enemies whose *raison d'être* was to destroy the lives of humans and animals implied that larger organisms should have defense mechanisms—warfare devices—to fight them. The concept of the immune system as the body's internal defense army may therefore have been derived from the "demonization" of microbes in the context of bacteriology. The medical accomplishments facilitated by bacteriology helped strengthen such military metaphors, and warfare became the dominant paradigm for disease and for medical intervention.

Since the late nineteenth century, the concept of disease as a result of microbial activity has dominated biomedical science.[5] This bacteriologic approach was used to control plagues, to distinguish and characterize particular diseases, to develop specific therapeutic drugs, to reduce human suffering and surgery-related infections, and to sanitize cities, open frontiers, and improve public health.

Germ theory has also liberated humans from the "blame the victim" models of Galenic medicine and the terror of contagious miasmas. Disease ceased to be about morals; it was now about external agents. Disease prevention was no longer achieved through prayer or by living a pure life, but by chasing out the evil microbes that had invaded our bodies. Morals and behavior become irrelevant compared to hygiene, sanitation, and antibiotics.

By reducing the scope of causality to the action of microscopic beings, germ theory influenced the bias of perception in disease causality by diminishing the role of the social sphere. Ironically, it is the human disease whose germ is associated with Koch,[6] tuberculosis, that best illustrates the limitations of the biomedical paradigm in the understanding of the prevalence of disease. To this day, tuberculosis remains endemic in correlation with poverty and other social variables. The presence of the *Mycobacterium,* or Koch bacillus, per se, does not determine the occurrence of disease.

The elusive search for a microbial cause for cancer in the 1970s was also framed by the germ-theory paradigm. Scientists of that time believed that if they could identify an infectious agent for cancer, they could also develop antimicrobial drugs that would make this disease curable (see Chapter 1). They did not find a cause for cancer in the 1970s, but, largely on the basis of their search for a cure, they identified a cause for AIDS in the 1980s, and named it HIV.

Yet, as seen in Chapter 2, waging "war" was not the only or the best way to arrive at meaningful results regarding AIDS. Author Bruce Nussbaum builds his argument in favor of community trials and the inventiveness of the PWA movement by putting it in opposition to the "war against AIDS"—in his view, a "war" lost by the "generals in charge," that is, the medical establishment.[7] It was not the generals but people outside those spheres who created the response to AIDS that seemed to work best. It was not the classic eradication method or the heavy-handed public health strategy of quarantine or *cordon sanitaire* that led to a hopeful response to this disease; instead, it was the subversion of such militaristic tactics. By establishing the possibility of living with AIDS and with HIV (see Chapter 2), some PWAs unknowingly advanced the terms of a new paradigm. By creating the PWA identity, they challenged the expectation of their death simply by continuing to live—by preventing additional infections, balancing nutrition, lobbying the biomedical community for more research and better medications—thus turning HIV infection into a chronic condition. In newsletters and through activist networks, PWAs developed a subculture of expertise in handling infections and promoting immunity that dialogues and overlaps with biomedical expertise, in a way that expanded the possibility of coexistence with a virus that resists contemporary medical warfare techniques.[8]

* * *

Before the onset of AIDS, Susan Sontag's reflections on the use of metaphor for cancer and tuberculosis became an influential best-seller.[9] This line of reflection was expanded and substantiated empirically in the field of AIDS.[10] Authors who analyzed the symbolic aspects of disease emphasized the hyperstigmatizing effect of metaphors in illness: once depicted as a battlefield, the body itself becomes the illness, the sick person becomes the disease, risk groups become the danger, and affected nations become the enemy. An overload of meaning feeds back discomfort, disorder, and more suffering. Sontag pleaded for a liberation from metaphors in disease and medicine. But how are we to replace the deeply ingrained germ theory that has framed our understanding of disease for a century and through which we have achieved so much in the field of public health?

A change of framework would, using terms given by Thomas Kuhn, be equivalent to a small scientific revolution, that is, a paradigm shift for the biomedical sciences.[11] What could replace the solidly established military metaphors of germ theory? Within the contemporary context of medical science, which includes the challenges of AIDS and a number of autoimmune disorders, we might think that significant changes should occur in the area of immunology. Since AIDS is defined primarily as a disease of the immune system and has mobilized major research efforts and funding, we might expect major advances in immunology through AIDS research. Acknowledging the vibrant character of that discipline, anthropologist Emily Martin asks whether that vitality is related to attention to AIDS or whether it existed independently.[12]

Having asked a similar question at the beginning of my project, I was brought to a different conclusion through my research. Even though AIDS has shaken the field of immunology, it does not seem to have caused a major conceptual transformation. Frameworks for investigating immune disorders remain unchanged in the busy and anxious field of AIDS research, even though some people involved with AIDS have come up with different concepts and models, including challenges to the warfare paradigm. It seems as though the majority of those involved with AIDS have been too busy saving lives or pursuing their careers to spare the time to consider a radical look beyond the main framework.

It was not within the field of AIDS, but very close to it, within the realm of basic science research, that I found the most radical questioning of the war paradigm. I heard the following statement from an immunologist:

> There is a dogma about anything in immunology being of defense . . . like military . . . reinforced by the old textbooks of immunology . . . where, for instance, lymphocytes are [represented as] little soldiers. . . . Doctors are trained under a paradigm where they are the good ones and the bugs are the evil ones . . . doctors must control a therapeutic weaponry, know it well and handle it well in order to kill the bugs. (Rodrigo, 50)

This statement seems to reflect the insight of a radical literary critic engaged in the deconstruction of scientific discourse, but that could not be further from the truth. It is actually an extract from an interview with two immunologists in one of the most prestigious research centers of South America: Instituto Oswaldo Cruz, Rio de Janeiro. They elaborated further on the inadequacy of military metaphors in medicine, infectious disease, and immunology:

> From a conceptual perspective . . . [the war between good doctors and their drugs against evil bugs] is absurd. What you have is a disadvantageous competition between transnational corporations, producing new drugs, and the microorganism, producing new resistances, which leads to nowhere except to cash profit for the corporations. It's absurd. You forget about health, about the healthy carrier, life in coexistence with bugs. (Rodrigo, 50)

If I were looking for a different paradigm, these statements would provide some clues to it. Their reference was something other than the model of antagonistic war, which they described as "absurd." Instead of demonizing the bugs and talking warfare, like the lay, mediatic version of immunology,[13] these immunologists argued for a broader understanding of the immune system as a complex system of interacting networks within which lies the ability to distinguish self and nonself, but not necessarily driven by a war system. This view is shared by a number of other immunologists, including some who were interviewed by Emily Martin at the Johns Hopkins Medical School.[14] In her anthropological exploration of the concepts of immunity—from high-ranking research laboratories to the streets of Baltimore to the mainstream media—Martin concluded that even though many immunologists think that warfare accurately portrays the functioning of the immune system, there is no consensus among scientists. The concept might better be described as a media favorite, one that does not even reverberate into the far more diverse opinions of the average layperson on the street.

Immunology is a dynamic field in which several competing models coexist. The "organism-centered" or network approach[15] has been expressed in mainstream publications such as *Immunology Today* and *Immunological Reviews,* but is not, so far, considered mainstream theory.[16] As a result of the overwhelming influence of molecular genetics on contemporary biology, complex systemic frameworks have more or less been suffocated, according to Brazilian immunologist Nelson Vaz.[17] They are not as popular as the overspecialized molecular way of thinking associated with the use of expensive high technology that gets most of the spotlight and funding. Far from the molecularized mainstream, "network" immunologists trace their understanding of self-regulatory systems to Metchnikoff and the origins of immunology, and often refer to the Australian immunologist MacFarlane Burnett.[18]

In their view, bugs and microbes should not automatically be considered "bad," or regarded as "enemies." One scientist I interviewed referred to a friend's project—"writing the history of pathology from the perspective of the microorganisms." Microbes do not have the teleological purpose to invade, assault, or kill the host organism; nor are they innately "evil," except in our human perception of them. Illness is just a particular interaction between microbe and host. Furthermore, the immune system does not exist just to act in the exceptional moments of their encounter. Actually, immune responses may not be triggered simply to distinguish self from non-self, which is the model more often presented as an alternative to the more militaristic one. Differences and contradictions exist within the self from the embryonic stage, as Metchnikoff demonstrated almost a century ago. The immune system acts in response to internal contradictions within the self that exist from its earliest stages.[19]

These immunologists pointed out that this inadequate and partial understanding of the relationship between the organism and the infectious agent underlies the persistence of war paradigms in contemporary medicine and immunology. That misunderstanding stems from the fact that most research has focused on the exceptional and intense phases of acute infection or transplant rejection, with a correlate disregard for immune function during non-acute phases.

> What is the immune system doing when you are not rejecting a transplant, when you are not fighting an "evil" bacteria? What is it doing then, why does it exist at all? [The problem is that] there is too much emphasis on [the study of] exception. Illness is the exception; we spend most of our lives healthy, interacting with all those bacteria, viruses, dusts—and we seem to manage all these "aliens" quite well. (Marie, 37)

My informants had a powerful insight into social metaphors and, like critical social scientists, they found the current model of the immune system as a human army that stands at the ready, waiting for an attack, to be inappropriate and anthropocentric. Based on the social experience of wars and armies, that model does not account for much biological activity, nor for the very existence of the immune system. This is not the only analyzed case of anthropocentrism in biology,[20] and dominant cultural themes of warfare are used cross-culturally to represent biological processes.[21]

Unlike social scientists, whose analysis consists mostly of drawing correlations between the metaphors used in biomedicine and external social variables such as gender, power, economic systems, and belief systems, the immunologists had a sense of what to look for instead of the perceived warfare. In their view, the immune system should not be studied by focusing on the intense and exceptional activity that goes on during periods of acute infection. The banal and unexceptional trivia of everyday life seemed to hold the key to the "revolutionary" shift in understanding. Resisting the study of the "special," another immunologist explained, "We are driven to study diseases . . . but my interest is about normal physiology . . . how the lymphocytes recognize the antigens, and how, in a general manner, they guarantee balance and health in the internal environment of the organism" (Jacques, 40).

It is not particular to immunologists and biomedical researchers to fragment knowledge and pursue details while losing sight of the big picture; there is a broader such tendency in cognition processes under certain circumstances. Detailed questions are easier to handle, and hyper-specialization is cultivated in our contemporary technologically oriented scientific culture—even if it is far more expensive and prevents less well-endowed research centers from making contributions. In the case of immunology, the pursuit of specific diseases and details on the infectious agents matches the fragmented character of the production of knowledge—to the extent that hardly anyone has a good understanding of how the immune system works outside the "feisty" periods when it responds to the presence of external bacteria.

Moreover, as many of my interviewees in AIDS clinical settings imagined, research in basic immunology could be even more complex and expensive than the already expensive study of the molecular structure of the virus.

The network approach, which is understood in terms of complex systems and chaos theory and which stems from Metchnikoff's theory of primal internal contradictions, is an attempt to draw a comprehensive picture of the immune system.[22] Even though the model seems too complex to be adopted in clinical situations, its implications for treatment may be surprisingly simple. Overcoming the molecular trend and the theme of specificity, "magic bullets" give way to generic therapies that can be as elementary and inexpensive as the oral ingestion of foods.[23]

* * *

The critique of the warfare metaphor in immunology from within an immunology laboratory was a novelty to me. I thought, as my experience with busy clinicians seemed to confirm, that scientists were not supposed to open their own "black boxes." They might speculate beyond germ theory, but since the advent of bacteriology they had framed their actions as if they were pursuing villainous disease-causing bugs. Their world view included identifiable, characterized diseases—for each disease a microbe, and for each microbe a drug or set of drugs. The increasing interest in genetics and the genetic co-determinants of many diseases did not dismiss the overwhelming power of germ theory, especially among infectious disease specialists.

A second look at the history of the Infectious Disease specialty, bacteriology, and germ theory may be of interest at this point, and may yield some clues about the centrality of warfare imagery in this field. Traditional accounts of the social history of bacteriology are associated with the findings of Pasteur and emphasize the impact of sterilization and its benefits to medical care and human health. However, germ theory had perhaps its most powerful impact in the then growing field of tropical medicine, which would develop into the specialty of Infectious Disease. The links are quite visible in the history of Brazil, whose most celebrated period corresponds to the "conquest" of tropical disease. Then, the warfare style merged with the imagery of breaking through a frontier, which fits so well into the mythology of scientific discovery. In other settings, similar links were established between colonial administrations and the development of knowledge about tropical/infectious diseases, in an intense symbiosis between the military and the medical endeavors.

Tropical diseases were seen as a fate of geography for vast regions of the globe, many of which were European colonies from the sixteenth to the nineteenth century. As a former colony that became independent in 1822, Brazil shared the burden of that stigma. The possibility of progress (the positivist motto adopted by the Brazilian Republic in 1889 and printed on the federation flag) was jeopardized by "tropical scourges." The internal frontier could not be developed and settled; its populations suffered heavily from a number of mostly unidentified forms of pestilence. The urban poor dwelled equally with plague and yellow fever. The fear of contamination scared Europeans and kept them from migrating to the country; it also had a negative impact on international trade. In this context, and in spite of the resistance of public agencies to supporting science, bacteriology became a central historical element of national development. In retrospect, it was represented not just as a scientific theory, but also as a liberating paradigm, a doctrine, a kind of lay religion that was to save Brazil from backwardness and allow for progress. The ports and cities and their pests, on one hand, and the inland frontier and its disenfranchised inhabitants, on the other, both had to be rescued from the ills that prevented progress—a difficult goal for a vast society that was just out of slavery and was highly stratified, with considerable socioeconomic inequities.

The conquest of infectious disease is remembered as one of the central sagas in Brazilian history. The doctors who helped fight endemic plagues are represented as civilizing heroes; their portraits hang on the walls of many research laboratories. They are part of a national iconography that transcends local history and represents Brazil's entry into the modern world, a process that on some levels continues to the present day.

The most celebrated of those pioneers was Oswaldo Cruz, after whom the leading South American research institute, located in Rio de Janeiro, is named. Historians describe him as quite isolated in his early devotion to the new and revolutionary bacteriological theories, with a drive that in 1896 led him to sail for Paris to specialize in microbiology at the Pasteur Institute. There he met Metchnikoff, the author of the theory of phagocytosis and a founder of the field of immunology, who became a close friend and invited him to work in Europe. Cruz chose to return to Rio in 1899 to "use there what he had learned in France."[24] He was able to put it into practice right away, for signs of the bubonic plague had appeared in several Brazilian harbors. First appointed to the Board of Hygiene, and later (in 1903) appointed head of the General Public Health Department, Cruz soon became involved with historic campaigns against yellow fever.

Local beliefs about yellow fever were based on miasma theory. People believed that proximity to the sick and their clothes caused contagion. Scientists tended to believe instead that some type of microorganism—*Bacillus icterioides* or *Cryptococcus xantofagus*—was responsible for the disease, but nothing could be proved. When a North American expedition visited Brazil in the late 1880s to do research on yellow fever, the disease was dormant, and they found nothing. However, by the time Cruz returned to Brazil, there was some evidence from Cuba that yellow fever was indeed caused by a microorganism. The Cuban doctor Carlos Finley (regarded by Cuban historians as the discoverer of the cause of yellow fever) had traced the transmission of yellow fever to mosquitoes. The American mission led by Walter Reed (regarded by American historians as the discoverer of the cause of yellow fever) conducted additional experiments, which simultaneously dismissed Finley's hypothesis and confirmed it by introducing the supplementary concepts of a microbial incubation period and a transmission cycle.[25] Inspired by the Cuban findings, Dr. Emílio Ribas, from São Paulo, published an article in January 1901 on "the mosquito considered as an agent for the propagation of yellow fever"[26] and coordinated the first Brazilian campaign against mosquitoes in Sorocaba, in the state of São Paulo. This occurred just before Walter Reed addressed the Pan-American Congress.

The campaigns against yellow fever in Rio had to mobilize information and propaganda in order to change people's beliefs and attitudes. Instead of burning the clothes of the infected, people were urged to protect themselves from mosquitoes and prevent them from breeding. Emílio Ribas jumped into the bed sheets of a yellow fever patient to show the public that there was no risk of contagion from clothing. Oswaldo Cruz was ridiculed in the media for chasing insects and was shown in caricature dueling with giant mosquitoes.[27] The success of the campaigns[28] strengthened Cruz's political power, but not enough to guarantee the approval of a bill making official the conversion of the serological institute of Manguinhos, where vaccine sera were produced, into a center for experimental medicine. The oligarchies considered it a waste of money, merchants opposed the state-supported manufacture of medicine, and the medical college opposed the teaching of medicine outside its own walls.

Only after Cruz won an international award in 1907 at the Hygiene and Demography Exhibition in Germany and became a national hero was the Institute of Manguinhos officially recognized. It was endowed as a research center and named the Instituto Oswaldo Cruz in 1908. By that time, its director had also built the exquisite Moorish castle that lodged the serological institute for many years, and which to this day stands out on the campus. Sparing no expense and luxury in its architectural detail and furnishings, Cruz is said to have wanted the castle to stand out over the *mangue,* the swamp above which it was located, "just as science stands out and shines over the ailments of the masses."[29] Yet another military evocation.

Structured as a research center, Manguinhos surpassed the Butantan Serological Institute of São Paulo, whose coordinator, Adolpho Lutz, later joined the team from Rio. The campus of Manguinhos was the site of the most pioneering research on infectious disease in Brazil, always articulated with major European and North American research centers.[30]

Also from Manguinhos came the coordination of scientific campaigns that were to go into the hinterland. Those expeditions were also consonant with the dominant paradigm of germ warfare. Their mission was to conquer, civilize, and annihilate microscopic enemies. By medicalizing their ills, they helped to remove the stigma of laziness formerly attributed to the inhabitants of the tropics. Bugs—not climate, laziness, or race—should be blamed for their illnesses. The imagery of those expeditions portrays young scientists (always white and male, and often from the urban elite) traveling in the wild to work among the afflicted (always poor, dark, and of any age or gender).[31] The scientists are represented as everything from miraculous saints bringing relief to the ailing masses, to civilizing heroes bringing redemption with their medical paraphernalia, to victorious warriors defeating the enemy, to triumphant hunters of evil microbes. The campaigns were taken to Minas Gerais, Mato Grosso, and the Amazon. One of the most celebrated findings was the identification in 1909 of the "American" disease trypanosomiasis, thereafter named Chagas disease in honor of Carlos Chagas. This same scientist is credited with every discovery related to the disease. He identified it among a mass of ailments and characterized its infectious agent, its vector of transmission, and its cycle of infection. Brazilian medical history celebrates not one discovery, but four—all made simultaneously by the same scientist.[32] Within that context came a discovery that would later prove to be of major importance: the protozoan initially named *Schizotrypanum* and described in 1909 by Chagas as a possible variant of *Trypanosoma cruci*. It affected the lung, and was later described in 1912 by Delanoe as *Pneumocystis carinii*.[33]

During the construction of the railroad in the inland state of Minas Gerais, Chagas was called in to implement a program of malaria prevention. He set up his headquarters in Lassange.[34] Reasoning by analogy to his knowledge of malaria,[35] he focused on the hemophile insect *Triatoma infestans,* which was quite widespread and was found on the uncoated walls of poor rural houses. This insect was known locally as *barbeiro* ("barber"), because it sucked blood from the human face. Chagas hypothesized that the insect was a carrier (vector) for a microorganism that caused the swollen necks, sweating, and malaria-like symptoms that, along with a heart condition, were symptoms of the disease.[36] This mode of transmission was similar to that of malaria and yellow fever, for which different species of mosquitoes are the vectors. Carlos Chagas isolated a *Trypanosoma* that differed from the one that causes sleeping sickness in Africa and named it *Trypanosoma cruzi,* in honor of Oswaldo Cruz. The *barbeiro* became the target for disinfestation and fumigation with DDT in order to prevent transmission, and treatments for the sick were developed.[37]

Throughout the 1920s, polemics persisted around the authenticity of this condition as a new and separate disease, as well as around the causative role of the newly identified agent. Those polemics seem reminiscent of today's fringe discussions about AIDS. They challenged the veracity of the evidence of a newly defined disease.[38] Only after other Latin American countries expressed their interest in the study of American trypanosomiasis did it become a topic worthy of study once again in Brazil.[39]

The field consolidated and has attracted many researchers up to the present day.[40] Protozoology, for instance, was developed significantly in Manguinhos as a result of the study of Chagas disease, and several lines of research on this pathology were implemented. The original research on the molecular interaction between the insect vector and the microorganism, based on the model provided by the interaction *Triatoma (barbeiro)–Trypanosoma,* served as a basis for exploring basic immunological questions that can be extrapolated for other infectious diseases with insect vectors, and that can also have direct applied effects. For instance, one group of researchers from the Instituto Oswaldo Cruz found that certain wild plants block the infectiousness of the *Trypanosoma* organism in barbeiro carriers, a finding that may have important consequences in the prevention of Chagas disease transmission.[41]

* * *

It is ironic that the first time I heard a consistent development of a radical critique of the germ-theory paradigm, it was within the specialty that was inspired by the toughest anti-microbe wars. This is not surprising, however, for in Brazil that field has remained a vibrant and fertile area of research. This particular finding matches the suggestion by the historian of Brazilian science Nancy Stepan that "a concentration on Brazilian problems did not rule out the possibility of making discoveries on disease mechanisms in general."[42] The vitality and universal value of biomedical research within the context of Brazilian tropical medicine has been documented by the historians of Casa Oswaldo Cruz.[43] The work of this group, a history and social science research unit within FIOCRUZ, is itself a sign of social interface with a local medical field. The strict unicausal theories of bacteriology gave way to more complex theories to account for host–parasite interactions and for the social variables involved in the determination of disease.

Another important site of knowledge production with interfaces with biomedicine is the National School of Public Health (ENSP), which is also part of FIOCRUZ. ENSP has produced works that combine social analysis and biomedical knowledge and has created a critical, creative atmosphere that constantly produces new knowledge on medical problems. A number of works conducted by local researchers explore the convergence of social and biological variables in the study of infectious diseases:

- Malaria, for instance, has been shown in correlation with the type of deforestation and land ownership practiced in the Amazon. The disease, which is caused by *Plasmodium falciparum* and other variants of a mosquito-transmitted microbe, affects each social stratum differently. The first settlers, who hold small parcels of land, generally succumb to the disease. The second wave of settlers sell the land at the first sign of malaria; by the time the land is suitable for settling, it has already been concentrated into large estates.[44]
- The prevalence of schistosomiasis is directly related to land tenure.
- Tuberculosis, dysentery, and a rise in child mortality continue to be directly correlated with poverty and undernourishment.
- Leptospirosis prevails in the city slums, where open sewers carry the urine of infected rats.
- Chagas becomes an urban disease through the transfusion of infected blood.

- Leishmania, which had been dormant for years and is considered a disease of the wild spaces, reappears in the perimeter surrounding households, both in urban slums and in the countryside.
- Cholera, which had been absent from Latin America for centuries, has been reintroduced via trade with countries in the Pacific Ocean. In the early 1990s, an epidemic of cholera decimated people in the Andean countries and threatened the inland population of Brazil.[45]

Overwhelming evidence of the social determinants of disease distribution makes its understanding a component of local medical culture. Rather than a question left for social scientists or public health workers, it is a part of medical knowledge, and more so of the specialty of infectious disease.

Local researchers have also pursued a number of exciting exploratory projects in microbiology. Studying the biology of "tropical" diseases, they have been able to address fundamental questions in biomedicine that have been left off the "molecular genetics bandwagon"[46] or escaped the influence of the "molecular imperialism" that has dominated most research programs during the last decade.

As infectious diseases remain major public health problems and are also acknowledged as social and development problems, the original alliance between pioneer researchers and the ailing masses remains. Attracted either by the potential for pure research, by social causes, by the specificity of the field, by its national tone, or by other reasons, a number of qualified researchers have been continuously attracted to the field of infectious diseases in Brazil. The local concentration of research questions and the pursuit of the complexity of factors regarding infectious diseases are exceptional in this context. Through dialogues with other research centers and some international networks, contemporary Brazilian scientists have been able to pursue aspects of infectious disease and immune function that may remain unexplored in the technology-driven centers of Europe and North America,[47] where interest in infectious disease has declined significantly[48] and molecular biology has become the hegemony of biological research.

* * *

In FIOCRUZ, in UFRJ, and in other Brazilian research institutions, exciting research took place in medicine, in the social sciences, and at the interfaces of medical and social questions. But let us not be caught in a "nationalist" point of view, nor in the Manichaean dualism that has been framed by both modernization and dependency theories. It is not "Third World knowledge" or a "Brazilian science" that emerges here. It is knowledge produced under the particular circumstances that I have attempted to describe—an idiosyncratic combination of overdeveloped, cosmopolitan sectors within an underdeveloped context. A strict definition of a Third World way of thinking and making science would lead us into the same kinds of conceptual traps that fostered efforts to dismiss scientific knowledge as bourgeois, male chauvinist, or western-imperialist. If essentialized, the privileges of the "partial perspective"[49] would become parochial, and the resulting alternative could become narrower instead of more comprehensive.

Most of my interviewees in clinical and biomedical research settings had a clear understanding of the dilemmas of being engaged in pioneering, top-level research problems while they were surrounded by the afflictions of underdevelopment. That structural tension appeared at all levels, from the allocation of budgets to the allocation of personal time, and from the choice of research problems to the choice of a sphere of action. Colorful descriptions of the structural tension abound: high-tech laboratories whose filters had broken because the children from the slum next door used the water reservoir to swim and play; a research facility equipped to use polymerase chain reaction (PCR)[50] in a city that lacks the equipment necessary to perform a simple necropsy to assess a cause of death;[51] the pursuit of theoretical research questions where even existing knowledge cannot be applied to save lives because basic essential equipment is lacking; the purchase of expensive antiretrovirals when simple antibiotics are lacking; the focus on AIDS where more people are dying of tuberculosis, malaria, and malnutrition; the use of satellite conferences for research updates when most of the people in the surrounding community can barely read.

This scenario is less a composite of world types than the face of the contemporary world—not the traditional geographic partitioning of developed and underdeveloped or, as was seen during European colonization, of metropolis vs. satellite, but a constant re-creation of centers and peripheries, flexible and mobile centers, and newly generated peripheries, all next door to one another.

In some cases, as I observed in Brazil, that proximity was incorporated into the production of knowledge and became central to the local production of biomedical science as well as to clinical practice. It is no coincidence that social epidemiology, social medicine, and the important field of infectious disease flourished in Brazil in close relationship to the classical endemic and epidemic diseases. The problems locally defined as research priorities are closer to becoming universal human problems than are those that may be defined in technology-oriented research settings.

It might seem, therefore, that AIDS had an ideal site for the local approach to science to make its contribution to global efforts. It seemed so at the time, when apparently genuine global efforts from international agencies and different partners conducted the international responses to the epidemic. However, AIDS research rapidly became a technology-oriented field, growing increasingly dependent on First World resources, strategies, and questions. "AIDS became like another cancer," remarked one of my interviewees, meaning that AIDS-related research became technology-dependent, adopting the strict models of biomedicine. "Only if the epidemic gets out of control, and does not get a vaccine or any classical treatment, might our [social] knowledge matter," said the same social epidemiologist, seasoned in his understanding of the broader social determination of disease.

Expressed in high-tech language and protocols and invested with emotion, anxiety, and exceptional funding, AIDS research rapidly evolved into technology-dependent forms of medical research. One physician commented,

> Even the diagnosis of tuberculosis [one of the most prevalent AIDS conditions] has now to be performed in the high-tech patterns defined in the first world, with expensive tests, when we always performed it based on clinical and differential diagnosis at no extra costs. . . . it seems wrong . . . we should be engaged in the development of more affordable means of diagnosis, that could be valid anywhere. (João, 35)

Some researchers mocked the "high-tech" style as being shortsighted:

> You know, sometimes [North] Americans demand the exact terms of proof they defined, otherwise they will not accept your data. If I send them the corpse, the end of it, with the picture of the corpse, it is obvious that it cannot be anything else than a corpse, and they will ask for this and this and that assay in order to prove that it is really the corpse, that it has the smell of the corpse, that it is chemically proven that it is the corpse. (Rodrigo, 50)

At other times, they saw it as a perversion of the system, fueled by the profit-making mentality of the corporations that produce technology:

> Who wants to see a simple-technology substance that could sterilize the blood in blood banks, for safe blood transfusion? Can you imagine the amount of money in serological test kits that would not be made by the manufacturers? (Rodrigo, 50)

While despised by some of my interviewees, the "high-tech mentality" was pursued by many others who concentrated their efforts on equipping laboratories with competitive testing and assessment devices. Researchers involved with AIDS, whether or not they were aware of it, may not have had any other choice but to adopt the idiom of high technology. This was the language in which all the new medical knowledge was expressed, and the only one through which their efforts could be accounted for. That same reason made them place themselves in a structurally subordinate position. They will never have the number of reagents or the kind of laboratory facilities, equipment, support staff, libraries, communications equipment, and money necessary to produce the level of competitive research that is acknowledged by mainstream medical publications.

> We should limit ourselves to research here things that they [the First World] will not research there, or have no interest in there. The type of things that will not lead to a Nobel, but can add to knowledge about AIDS, and that we can produce here safely. (André, 37)

Some interviewees had a clear idea that the research funds allocated by international agencies to Brazil covered only clinical descriptions, epidemiology, and eventually the social aspects and local characterization of viral varieties, but not major influential basic research. Some acquiesced to that order of things, which found powerful support in the technological argument: How could Brazil compete otherwise? There is no money, no resources, no infrastructure.

Another group among my interviewees in basic science despised that type of conformity. They insisted that scientific creativity has nothing to do with the level of technology available, and that the most needed resource—the patient with the infections—was fairly abundant in Brazil, certainly more so than in the northern countries. Mystification by high technology separated the waters between the two types of researchers. And, as far as I could observe, research on AIDS was absorbed by the more technology-oriented style, in Brazil as elsewhere. As a consequence, the "original" contributions—inspired by Chagas-style research—fell behind in the "global" efforts to respond to AIDS.

However, the scientific creativity that heightens the possibility of a paradigm change in immunology or in the understanding of the social determinants of diseases is alive and well and existing side by side with AIDS research. Maybe sometime in the future it will contribute to broaden the models of understanding disease that are central to medicine.

SEVEN

Conclusions

HAVING APPROACHED THE FIELD of knowledge about AIDS from several perspectives, we can now evaluate how the ethnographic data and analysis address the questions raised in Chapter 1 concerning the social construction of scientific knowledge and the structures of asymmetry in the contemporary world.

In the account of the AIDS social movement and treatment activism in the United States (Chapter 2), I analyzed a process that we may now call the unleashing of knowledge making. That process made transparent to the public what used to be the opaque walls of research; to use a favorite metaphor of science analysts, we opened wide the "black boxes" of science.[1] They were not deconstructed and torn apart by the usual academic critics, who, for this purpose, came relatively late. The force that broke into the "ivory towers" of science came from society itself, or, to be more precise, from a segment of society that had never been involved either with science or with its analysis: those who were directly affected by the disease.

With their lives at stake, AIDS activists scrutinized the reasons why there was no cure for them in a century whose claim to fame includes medical victories over infectious disease. Initially they blamed the government's and scientists' lack of interest in their lives. Since most of those affected by AIDS were gay, activists pinpointed homophobia and prejudice as the underlying reasons for government inaction. Once scientists became motivated to conduct research on AIDS, and government funds as well as private money were raised, activists selected other targets for blame: bureaucratic inertia, corporate greed, and the personal vanity of individual scientists motivated by grants, awards, and publication rather than by the prospect of improving public health and saving human lives. Equally fueled by anger and filled with indignation were some of the journalistic reports on AIDS. They depicted greedy scientists and businessmen conspiring behind closed doors to establish the profits and awards that AIDS might bring them, while disregarding the needs of those affected by the epidemic.

When the number of bureaucratic obstacles was reduced by political decisions, and when scientists as well as laboratories became publicly accountable, activists realized that the problem was more profound: the lack of scientific imagination and experimental daring constituted as powerful a limitation to the achievement of new knowledge as the will of scientists and governments.

Meanwhile, activists unleashed the process of knowledge making in several directions. They developed their own community research initiatives. They fought and negotiated with public agencies and with pharmaceutical corporations. They made recommendations and sat on advisory committees. They opened the previously closed doors of scientific research. The process of science making trickled down to society. Knowledge started to be produced outside laboratories, too. Negotiations between scientists and society defined the standards of consensus about critical "scientific" issues, such as clinical trials and drug approval.

As social activists became co-producers of knowledge and shared the responsibilities of research and its results, their attitude toward the "black boxes" of science moved from anger to disenchantment. As research doors opened—as a result either of forceful pushing from the outside or of negotiated agreement between the parties involved—and as the boundaries between outsiders and insiders became blurred in the process of negotiating the terms of action and defining research questions, the depictions of oblivious scientists unwilling to engage gave place to a wider disenchantment about the current limitations of science. This is where the subject begins for those who are engaged in the social study of scientific knowledge. Are those limitations and our understanding of them a sign that things are about to change, and that we may be in a period of Kuhnian revolutionary science?

* * *

From outside the mainstream ways of science, beyond the wider circle that involves those who protest at the agencies' doors and those who negotiate with them, some voices of radical dissent in the United States claimed that the basic assumptions of current knowledge about AIDS were wrong.[2] From one perspective, these views seemed dangerously irresponsible: they challenged the very basis on which every single tool that could be used against AIDS is derived. If those were questioned, what would be left? Relativism is uncomfortable in matters of life and death. And when our health is at stake, most of us prefer to follow un-deconstructed prescriptions. Seen from afar, however, those voices were merely demands for a change in knowledge models. They might have been inadequately articulated, but they reminded us that knowledge results from a constant process of construction and change, and that knowledge about AIDS has been particularly fragile and volatile. Examples of certainties being rapidly replaced by their opposite or by very different ones abound, including in the various sub-fields of AIDS knowledge—from etiology and pathogenesis to treatment, epidemiology, prevention, and social policies. What was thought to be a gay cancer became an infectious disease. What was initially related to modern decadence became associated with primal infections originating, like humankind itself, in equatorial Africa. What was seen as a disease of the urban metropolis and developed countries became a Third World burden, affecting the disenfranchised peoples from Africa, Asia, and Latin America. What was associated with anal sex became linked to pregnancy and birth. What was seen as an inevitable killer disease became in some cases a treatable chronic condition. What was seen as the only medicine, AZT, was discarded as single therapy for its limited efficacy. What was seen as being related to environmental and lifestyle issues was later attributed to a virus. What was seen as having only a viral etiology became recognized as having multiple causative factors.

What constitutes official and acceptable knowledge in AIDS changes constantly, mirroring the very nature of science making. Why do some ideas remain while others are rejected or revised? In the logic of scientists, ideas are replaced when there is either evidence against them or a new model that better explains the phenomenon in question. The acceptance of a new model is a matter of consensus, an unquestioned process for scientists and yet, from the social science perspective, the center of the problem: What are the social variables implied in that process of legitimization?

With the data presented in Chapter 2, it is possible to argue that the AIDS social movement in the United States introduced innovation into the process of consensus production. The same social actors that opened the doors of laboratories became part of the process. They influenced the criteria for clinical trials, trial design, drug approval, AIDS case definition, prevention strategies, and, later on, research questions in basic science. What in the past had been decided by communities and networks of scientists was now also discussed by those directly involved in the issue.

This fact challenges the late 1970s and 1980s wisdom of social constructionism and ethnographies of laboratory.[3] Within the context of AIDS, scientific knowledge is not just what is produced behind laboratory doors, seeking grants and peer recognition, measured by the number of publications and acknowledged by awards. No longer exclusively composed by networks of peers driven by academic rules or subject to the pressure of profit-making corporations and state policies, the social basis for the negotiation of consensus was expanded to the community of people whose motivation was their own survival. Did this make the knowledge produced less "relative," or "arbitrary"? Although we could describe the new knowledge produced as revolutionary and innovative, we cannot conclude that it is truer and more absolute than what was produced in the traditional form; we can only say, in a "quantitative" understanding of the word, that this knowledge is more "universal," or extended to more people. We may also have found an example for discussing the question of the social conditions of "scientific revolutions," a matter that Kuhn left to be pursued by others.[4]

* * *

My initial questions were not limited to the social construction of science, but extended to its relationship with contemporary world history. Roughly at the same time that activists in the United States initiated the process of unleashing science making, international agencies such as the World Health Organization (WHO) envisioned the global dimensions of AIDS and pushed forward an agenda for global action. Did that "global" effort correspond to a search for universality? Would it have an effect on the production of knowledge and make it even more "universal"? Was there a possibility of a generalized "global unleashing" of knowledge making? Such a scenario raised further questions about the plurality and multi-vocal nature of contemporary culture and the debates on modernity/post-modernity, which AIDS also embodied. Unlike the period of history known as the Cold War, there is not much of a consensus in the characterization of current configurations of world politics and culture. With regard to our subject—the social production of knowledge about AIDS—the world has been idealized as a global partnership engaged in the pursuit of a common goal. However, it may also be presented as a complex, wary, and sometimes conservative universe of dissonant voices, as some of the UN-sponsored global forums have shown. What channels of communication cut across that diversity and imminent cacophony? Who are the brokers in the process, and where does their social legitimacy come from?

If the social bridge that unleashed the process of science making within the context of AIDS had originally been between an empowered social movement in the developed countries and its science agencies, what type of bridge should be expected between the centers of knowledge production and the vastly disenfranchised populations that make up the majority in this world? The global agenda of WHO, their traditional advocate, accounted for local empowerment and full participation for all. In Chapter 3, we analyzed the role of WHO and other global agencies in spreading around the world the tools and models needed to act as fast and as efficiently as possible, not just in strict biomedical terms, but by promoting community empowerment as well. Was that agenda inclusive enough to account for a widely expanded basis for the production of scientific knowledge, to promote bilateral partnerships between peripheral and central scientific settings? Could the distant voices from the Third World be incorporated into the mainstream process of consensus making? To answer those questions, it was necessary to conduct empirical research, preferably in a developing country.

* * *

From the very beginning, local actions against AIDS in developing and peripheral countries had a source of strength in international agencies such as WHO, which funded them and helped them take off. Except for New York City and California, which staged original social responses to the epidemic, action in most sites can be analyzed only as a combination of local characteristics and international aid. In Chapter 4, I analyzed in detail the unique characteristics of one particular setting: Rio de Janeiro, Brazil.

The rich social movement responding to AIDS in Brazil from 1985 to 1995 cannot be understood without reference to the international movement and to the historical moment lived then by Brazilian society, which was overcoming two decades of military rule. Diverse and heterogeneous, the Brazilian AIDS–NGO movement tried unsuccessfully to create organized networks and define common agendas for national intervention. From within that diversity, the most audible voices fought for generic civil rights and opposed the government systematically. In the post-military transition, the relationship between the government and the society did not flow easily, and action on AIDS was no exception. While many North American activists worked under the assumption that the government represented them and should execute their demands as citizens, Brazilians distrusted their government, assuming that it was authoritarian, corrupt, and against the interests of the people. That radical opposition, matured under decades of authoritarian rule, was stronger than the availability for communicative action and interchange leading to any form of consensual knowledge; the universe of knowledge within which activists moved was irreducible to that of the government, and of the medical establishment, and vice versa. Under these circumstances, the initial wave of activism in Brazil created and became specialized in its own channels of knowledge making. This production included a wide variety of media interventions and academic style of social research. Some of the local production on social issues was acknowledged and adopted by international agencies.

* * *

In the medical settings, analyzed in Chapter 5, the battles were of another kind, and mainstream knowledge had to be handled on a daily basis. Because of the pressure of providing clinical care, there was not much time or energy to pursue original research questions, even though there were many aspects of the disease in the local setting that triggered speculation and raised questions. There were local infections and parasitic diseases that were different from those described in the literature, there were specific patterns of nutrition (the health care system had its own form), there were local belief systems that had an effect on how symptoms came about and how they were to be managed, there may have been idiosyncratic epidemic patterns due to local ways of transmission, and there were specific local political conditions. There was an intellectual awareness of these issues, but the possibility of turning them into research questions was limited. With few exceptions, they could not be converted into scientific knowledge that would be accepted by the international system.

One exception was the inclusion of tuberculosis among AIDS pathologies. Tuberculosis had initially been identified by Brazilian doctors as a possible opportunistic infection for AIDS.[5] In 1985, that hypothesis was dismissed by the Centers for Disease Control (CDC) as irrelevant.[6] At that time, tuberculosis was not seen as a public health problem in the United States because of its temporary invisibility and the cutting of funds for TB surveillance. Right afterward, the disease came back to U.S. inner cities and soon became a general problem, with serious implications related to multi-resistant strains. At that time the expertise of Brazilian doctors in the treatment of tuberculosis, which had long been considered irrelevant in developed countries, became important to everybody.

Another exception corresponds to the adjustments in the AIDS case definition made by Brazilian and other South American health authorities. The revised versions, which take local circumstances into account, were incorporated by international agencies.

These two examples, however, do not amount to a general transformation in the social basis of science making. Even though local AIDS experts were fairly cosmopolitan and visible, their participation in the world of knowledge production on the international scene was limited to sharing consensual knowledge rather than negotiating it. They had full and rapid access to international knowledge, but not necessarily in an interactive mode. Knowledge followed the traditional one-way, center-to-periphery route.

As the canon of scientific research is defined by international, anglophone journals such as the *New England Journal of Medicine, Annals,* or *Lancet,* so are the relevant questions and the relevant research. This fact leaves out the problems that might be raised regarding specific clinical problems that are found in Brazil. Such problems do not develop into fully formed research problems, particularly when the local actors are not trained to think of themselves as researchers, and have not mastered the idiom of research, or the know-how to develop original protocols, submit results, and get them published. Participation in international research is therefore limited to joint ventures in which the research questions and methods are defined by the First World partner. These protocols often involve the latest technology and expensive equipment, which itself promotes dependency and local differentiation.

Unlike the empowered AIDS activists in the United States, local doctors who treated people with AIDS were not in a position to develop an original perspective on the subject. Local activists had done so for social issues, and at some level they addressed the question in epidemiology as well. However, clinical and basic research problems were rarely pursued by local social actors, unless their terms were predefined. There was no "global unleashing" of the knowledge-making process. Knowledge flowed throughout the world, but through limited channels and rarely in a multidirectional manner. Persuasion and hegemony, rather than a truly interactive dialogue, characterize the development and transmission of knowledge. Local creativity is used to develop solutions to local clinical problems and is rarely translated into knowledge that can affect mainstream science. Therefore, generalized participation in the fight against AIDS will not necessarily bring about a significant shift in medical knowledge, which will probably tend to follow the same paradigm and function as "normal science."

* * *

Data presented before Chapter 6 provide a picture of dependent knowledge making, in which "global efforts" are mainly rhetoric, one that succumbs to the dependency mechanisms of the use of high technology in research. However, I had the chance to move beyond the field of AIDS research and investigate other aspects of the rich tradition which made me choose Brazil as research site: to explore the field of immunology as it developed within tropical medicine, and the field of social epidemiology as it developed from the analysis of health in dependent economies. Very close to AIDS, these two fields maintained original and alternative research tracks.

Social epidemiologists, who created complex models to account for the social determination of disease, remained distant from the AIDS research bandwagon, where the relevant variables are defined in terms of individual behavior. Mathematical models and statistical projections based on those variables are more easily used to make immediate decisions regarding public policies, and that is mainly what epidemiologists working directly with AIDS—including those in Brazil—have done. Thus, local epidemiologists who were directly involved in AIDS became dissociated from local traditions, and the bridges between them and their clinical surroundings were hard to create. Only at a later time, and mostly as a result of their review of epidemic patterns in Africa, were some elements of social epidemiology included in the overall understanding of this pandemic.

As for basic medical research, scientists pursued original questions on the immune system, the physiology of lymphocytes, or interactions between cell and virus that could result in a useful model for the study of AIDS. As illustrated in Chapter 6, they tried to provide an alternative to the warfare model that characterized mainstream research projects on AIDS and immune function. They emphasized the complex system network model, which might shed a completely new light on the way AIDS and other immune disorders are investigated in laboratory settings. It could have, it might have, but it didn't.

Why was the network model not incorporated into basic research on AIDS? Maybe the time has not yet come. If there was any link between the ideologies that characterized the spirit of the times and the models that framed scientists' metaphors and language, it would be found in warfare. Even if immunology is ready to replace its warfare paradigm with the cognitive network and flexible system imageries, it seems as though society may still be too comfortable with warfare.

Through the course of the AIDS epidemic, we seemed to be moving out of a long historical period defined by war, hot and cold, into a middle-range and centerless state of flexible accumulation and massified globalization. As we progress through the 1990s, however, the mirage of a post–Cold War era of openness, dialogue, and peace has succumbed to increasing violence, scattered wars, ethnic cleansing, and religious intolerance. Maybe science has plenty of models to replace the war paradigm, but society is not ready for that shift.

Notes

1. AIDS and Science

1. See WHO (1979, 1981, 1986a). For more details, see Chapter 3.

2. See, for example, Knowles (1977), American Academy (1986), and particularly Thomas (1977), Fredrickson (1977), and Rogers (1986).

3. See Rogers (1986).

4. See Chubin and Studer (1978), Fredrickson (1977), Studer and Chubin (1980); see also Harden and Rodrigues (1993).

5. War may be seen as one of several possibilities of dual oppositions, which themselves constitute the basic structure around which human thought processes—and, for that matter, culture—are organized, at least according to the structural anthropology inspired by Lévi-Strauss (1962a). Yet there are asymmetrical values of good and evil attached to the parties at war; war is too invested with feelings, passion, and morals to be just an ordinary form of opposition.

6. See Brandt (1987) for more information on infectious diseases.

7. See Stuber and Chubin (1980). Later, cancer research would jump on the "molecular biology bandwagon" (Fujimura 1988).

8. Gallo (1991), Lapierre (1990), Nussbaum (1990).

9. Garrett (1994).

10. In the second half of the 1970s, a few retroviruses were identified in animals. They are now known to exist in cats, sheep, bovines, and several types of primates. Today, the retrovirus is not the simplest infectious agent known. Some proteins can be infectious—for example, the *prion,* a particle associated with spongiform encephalopathy.

11. Retroviruses contain a single-chain nucleic acid (RNA), not the double-chain nucleic acid (DNA) found in regular viruses in the form of a double helix (the two chains twist around each other at regular intervals). The chains in both types of nucleic acid are made up of specific sequences of nucleotide bases, and the sequence of bases determines the genes (a discrete segment of nucleic acid) that the nucleic acid contains. Genes contain a code that tells the cell to make specific proteins, which carry out the cell's functions. Genes do not work in a virus, because the virus does not contain the equipment needed to express a gene (that is, to "turn" a gene into a protein). Once a virus deposits its nucleic acid into the cell, the genes can function. Viral DNA can tell the cell what to do directly. RNA has to become part of the host's DNA; this requires the action of the enzyme reverse transcriptase. Once part of the host's DNA, the code contained in the viral RNA has to wait until the right host genes are turned on before more viral RNA is made, as well as the proteins needed to make the "coat" that surrounded viral nucleic acid. Sometimes "mistakes" are made, which result in the production of abnormal viral RNA or viral proteins. The host can still make the products necessary for retroviruses, but the "new" retrovirus doesn't always work the way the original retrovirus did. For example, it might not succumb to the effects of drugs that the patient has been taking and, thus, is rendered drug-resistant—and very dangerous.

12. See Fettner (1990), Jaret (1994), Watson and Crick (1953).

13. Until the mid-1970s, biologists believed that DNA served as a template for the production of RNA, and that RNA, once it was released, did not become part of DNA ever again. Retroviruses contradict that dogma, in that their RNA becomes incorporated into the host's DNA, through the process of "inverse replication." This process is made possible by the enzyme reverse transcriptase. Howard Temin and David Baltimore were awarded the Nobel Prize in 1975 for discovering this process. In an interesting testimony on how science works, virologist Emmanuel Heller, from Tel Aviv, reported that he had observed inverse replication in the early 1970s but did not dare present his findings for fear of challenging the dogma (interview, Amsterdam, 1992).

14. Later, Kaposi's sarcoma was redefined as a sexually transmitted infectious disease rather than a cancer of obscure etiology. KS classification is still subject to discussion as we finish this manuscript.

15. Mass (1990a: 132).

16. Norman (1986: 170).

17. Shilts (1987).

18. Journalist Randy Shilts quotes Robert Gallo's remark that the editor of *Science* "thought RUB was a disgusting acronym for a virus relating to this particular disease" (Shilts 1987: 264).

19. After years of debating whether the virus was first discovered by the French or the Americans, the scientific community began to wonder whether the irregularities identified in the American laboratory were due to accidental contamination or conscious behavior on the part of members of Gallo's team.

20. Shilts (1987: 452).

21. Shilts (1987: 450–451).

22. Nussbaum (1990: 37–38).

23. Montagnier and Gallo (1989).

24. Whether the "epidemiological transition" (see Weindling [1992]) was due to medical advances or to improvement in nutrition standards (see McKeown [1976], Vogel and Rosenberg [1979]), it can be interpreted in different ways. Most often it is seen as the clinical counterpart of the demographic transition—that is, as a step along a modernizing evolutionary path, like Rostow's (1960) "take-off." Under the more critical view of "dependency theory" (see Amin [1976], Cardoso and Faletto [1967], Frank [1967], Wallerstein [1974–1980]), which states that underdevelopment does not correspond to literal backwardness but rather to an effect of the system of asymmetries that derives from an unequal exchange, the contrasts in morbidity and mortality patterns at an international level should be understood as a consequence of economic dependency, the aftermath of colonialism, the impact of aggressive plantation systems, and other socioeconomic factors (see, for example, Arnold [1988], Minayo et al. [1995], Navarro [1981]).

25. Earickson (1990), Mann et al. (1992), Panos (1989, 1990), Sabatier (1988), Schopper (1990), WHO (1988c, 1989b, 1989c, 1992, 1994b).

26. See, for example, Amat-Roze (1993), Crawford (1994), Farmer et al. (1993), Feldman (1994), Perrow and Guillén (1990), Singer (1994b), WHO (1994b).

27. See Lindenbaum (1992).

28. After this manuscript was completed, a lengthy comprehensive analysis of the negotiating process—*Impure Science*—was published (Epstein 1996).

29. See, for example, Crimp (1990), Feinberg (1994), Kramer (1989), McKenzie (1991), Nelkin et al. (1991).

30. See Burkett (1995), Garret (1993), Kinsella (1989).

31. Arenas (1993), Callen (1987, 1990), Crimp (1987), Daniel (1989), Feinberg (1994), Guibert (1990, 1991, 1992), Kramer (1989), Monette (1988), Schecter (1990), Watney (1994), Wojnarowicz (1991).

32. Fleming et al. (1988), Mann et al. (1988), Mann et al. (1992), PAHO (1989), Panos (1989, 1990), Sabatier (1988), WHO (1988b, 1989b, 1994b). See Chapter 4 for a discussion of the founding and activity of the Associação Brasileira Interdisciplinar de AIDS (ABIA).

33. Mann et al. (1992), WHO (1992).

34. Bolton (1989), Feldman and Johnson (1986), Horton (1989), Horton and Aggleton (1989), Pollak (1988), Pollak et al. (1986), Watney (1990, 1994).

35. See, for example, Bolton and Singer (1992), Bowser (1994), Davies et al. (1993), Gorman (1986), Gupta and Weiss (1993), Hahn (1991), Koester (1994), Magaña (1991), Pivnick (1993), Pollack (1994), Schiller (1993), Sobo (1993).

36. Aggleton, Hart, and Davies (1989, 1991), Aggleton, Davies, and Hart (1990, 1992, 1993), Berridge and Strong (1991, 1993), Carter and Watney (1989), Farmer, Lindenbaum, and Good (1993), Fee and Fox (1988, 1992), Graubard (1990), Herdt and Lindenbaum (1992), Loyola (1994), Paiva (1992), Parker, Bastos, et al. (1994), Thiandiére (1992).

37. Bayer (1985), Bolton (1995), Carrier and Magaña (1991), Davies et al. (1993), Gorman (1991), Katarbe and Lang (1986), Herdt et al. (1990, 1991), Murray (1990), Pollak (1988), Pollak and Schiltz (1987).

38. Carlson et al. (1994), Clatts et al. (1994), Connors (1992), Friedman et al. (1989), Goldsmith and Friedman (1991), Kane (1991), Neaigus (1994), Neaigus et al. (1990), Ratner (1993), Siegel (1988), Stall and Wiley (1988), Sterk-Elifson (1993).

39. Ankrah et al. (1994), Clatts (1993), Goldstein (1994), Gupta and Weiss (1993), Henderson (1991), Schoepf (1992a), Sobo (1993), Tuchel and Feldman (1993), Ward (1993), Worth (1990).

40. Caldwell et al. (1992), C. Good (1995), Packard and Epstein (1991), Schoepf (1991, 1992a, 1992b).

41. Farmer (1992).

42. Parker (1987, 1988, 1991a, 1992, 1994a).

43. Bolton (1992), Bowser (1994), Clatts (1991), Lewis and Watters (1989), Lindenbaum et al. (1993), Quimby (1992), Singer (1994b), Singer et al. (1994), Sterk (1993).

44. Bolton (1992), Cagnon (1988, 1990), Carrier and Magaña (1992), Davies and Coxon (1990), Parker (1988, 1992).

45. Crawford (1994), Farmer (1994), Frankenberg (1994), Gilman (1988), Goldin (1994), Héritier-Augé (1992), Pierret (1992), Schiller et al. (1994), Threichler (1988, 1992).

46. Abramson (1992), Lindenbaum (1992), Pollak et al. (1992), Threichler (1991).

47. Bateson and Goldsby (1988), Scheper-Hughes (1994).

48. See AAA (1993), Singer (1994a), Scheper-Hughes (1994).

49. See Leslie (1990).

50. Hannerz (1988).

51. See Habermas (1981, 1983).

52. See Burkett (1995), Camargo (1994), Moraes and Carrara (1985a, 1985b), Center for Social Research (1991), Epstein (1996), Nussbaum (1990), Patton (1985, 1990), Threichler (1987, 1988), Root-Bernstein (1993). Nonetheless, science is the unifying force that defines the ills experienced by a woman in Haiti, a gay man in Manhattan, or a child in Zaire as one and the same—that is to say, an HIV infection or AIDS. Likewise, a laboratory assistant running a PCR test, a statistician modeling projections on HIV for the next decade, and an outreach social worker interviewing prostitutes and housewives share the same topic of research, unified by science on AIDS. Moreover, a new and transnational identity is defined by the "AIDS doctor," who crosses over different specialties, locations, and backgrounds. Added to this new profession are the "AIDS specialists," who make up the "AIDS establishment," which, as diverse as it may be, is still defined under the unifying criteria dictated by science.

53. For the concept, see, for example, Crick (1982), Elias (1971), Kulick (1983).

54. See, for example, Mulkay (1979: 1).

55. The equivalent debate has been addressed by medical anthropologists, who have discussed whether biomedicine should be studied as just one more ethnomedicine, the one produced by western modern cultures, or as a special and legitimate scientific field that is free of cultural constraints. The debate has been settled through mature studies in the anthropology of biomedicine that do not slip into relativism. (See Good [1994], M. Good [1995], Greenwood et al. [1988], Hahn and Gaynes [1985], Lindenbaum and Lock [1993], Lock and Gordon [1988], Stein [1990].)

56. See Mulkay (1979: 2, 26).

57. Bourdieu (1975: 40).

58. Bourdieu (1975: 40).

59. See, for example, Barber (1990), Barnes (1974, 1982, 1988), Barnes and Shapin (1979), Barnes and Edge (1982), Bloor (1971, 1976, 1983), Brante et al. (1993), Chubin and Restivo (1983), Collins (1982, 1983, 1985), Fuller (1993), Hess (1992, 1995), Hicks and Potter (1991), Jasanoff et al. (1995), Knorr-Cetina and Mulkay (1983), Labinger (1995), Pickerin (1992), Pinch (1992), Restivo (1995), Shapin (1992), Traweek (1993).

60. See, for example, Arnold (1993), Good (1994), Wright and Treacher (1982).

61. See, for example, Fleck (1935), Keller (1995), Lewontin (1992), Reif (1961).

62. Latour (1990).

63. Haraway (1991), Keller (1983, 1985), Labinger (1995), Löwy (1989, 1990, 1991, 1995).

64. See, for example, Harwood (1986), Freudenthal and Löwy (1988), Löwy (n.d.).

65. Merton (1973).

66. Latour (1983: 204–205).

67. Weber (1963).

68. See, for example, Durkheim (1912). One stem of the sociology of knowledge comes from the rationalist Durkheimian tradition. In *Elementary Forms of Religious Life,* Durkheim (1912) described religion as the apotheosis of society and showed how this primitive form of thinking was the basis from which others, including science, were derived. Knowledge was seen as being shaped fundamentally by social categories that preceded logical categories, as stated in the founding essay on primitive classifications (Durkheim and Mauss [1963]). This tradition, which influenced

most of the social science literature written in French or about issues related to France, relied on faulty philosophical arguments (see Crapanzano [1995], Mulkay [1972]) and did not provide a firm basis for the development of a sociology that could address the contents of science.

69. See, for example, Marx (1973). Marx's philosophy demonstrates that systems of ideas are socially determined, even though the status of scientific theories within a multitude of other ideologies remains less explicit. Throughout the twentieth century, Marxist authors and politicians evoked the distinction between proletarian and bourgeois science in a way similar to the treatment of art in the early years of the Soviet Union—without, however, exploring much of its content. Marxist theorists, from Lenin to Althusser, distinguished "truthful" scientific knowledge from "untruthful" ideological knowledge. While this distinction matters for social scientists, it hardly works in domains such as biology, physics, and mathematics (for the rare attempts at applying it to these fields, see Baracca and Rossi [1976], Barletta [1978], Levidow [1986], and the classical Hessen [1931]).

70. See, for example, Mannheim (1936). Unlike the "ideology = false representation" supported by most Marxist theorists, Mannheim's ideology–utopia duality established creative principles for a future sociology of knowledge, which was developed only a few decades later.

71. Another anthropological perspective addressed science in relation to magic. Evolutionists considered magic a form of primitive science which, although permeated by irrational beliefs, established relationships between facts (Frazer [1900], Tylor [1870]). Functionalists demonstrated that magic, science, and religion were parallel, rational systems of action and thought that allowed humans to understand nature (Malinowsky [1948]). Structuralists elaborated distinctions between the internal logic of the two systems (Lévi-Strauss [1962a]) and showed how the relationships between humans and various aspects of the natural world expressed in totemic religions are actually about social relations among humans (Lévi-Strauss [1962b]).

72. Berger and Luckman (1966).

73. Kuhn (1962).

74. Coinciding with the upheaval of science studies, and certainly for parallel reasons, anthropology experienced its own moment of introspective reflection (see, for example, Marcus and Fischer [1986]). Following the post-colonial proximity between ethnographer and the subjects of study, disturbing questions arose regarding the accuracy of what used to be taken for granted from ethnographers' testimony. Focusing on the character of anthropologist-as-author, this current briefly disturbed the discipline, leaving positivists and realists threatened by the specter of hermeneutic conversion and the loss of scientific virtue. The movement helped to relativize the authority of ethnographic knowledge by analyzing the social conditions under which it was produced. In that respect, it should be seen in association with efforts to analyze the social production of the natural sciences and technology.

75. Collins (1985), Gilbert and Mulkay (1984), Knorr-Cetina (1981), Latour and Woolgar (1979), Lynch (1985), Traweek (1988).

76. Keller (1985), Fausto-Sterling (1985), Haraway (1989), Harding (1986, 1993), Harding and O'Barr (1987), Jacobus et al. (1990), McNeill and Franklin (1991), Martin (1987).

77. See Franklin (1995b).

78. See Rouse (1987: xi).

79. Brown (1989), Gross and Levitt (1994).
80. Reid (1975).
81. Rubinstein et al. (1984), Turnbull et al. (1989).
82. See Hess (1995) and Harding (1998).
83. Center for Social Research (1991), Marcus (1995), Penley and Ross (1991).
84. Escobar (1994).
85. Forsythe (1993).
86. Edwards et al. (1993), Franklin (1995a), Ginsburg and Rapp (1995), Longino (1995), Rapp (1993), Strathern (1992).
87. Heath and Rabinow (1993), Hilgartner (1995), Rapp (1994).
88. Clarke and Fujimura (1992), Fujimura (1987, 1988).
89. Haraway (1989), Martin (1992, 1993, 1994).
90. Epstein (1991), Erni (1994), Horton (1989).
91. Dumit (1995), Rabinow (1993a, 1993b).
92. In recent years, the "science wars" (Ross [1996]) arose out of a few disagreements that had been brewing between some scientists and some analysts of science. Suspicious of the anti-science attitude that radical relativism may generate, some scientists and rationalist philosophers dismissed the endeavor of Science & Technology Studies. Social analysts were not forgiven by "hard-line realists" for daring to cross the line, which Merton had so strictly respected, and "intrude" on the contents of science. The backlash against SSK (Restivo [1995]), or "STS bashing" (Hess [1995]), increases the challenge for social researchers in this unsettled field. This is even more so when the subject is medicine, because matters of life and death depend so much on it and because, unlike the fields of nuclear physics or biochemistry, clinical researchers face these matters on a personal and daily basis. For discussions that preceded the height of the "science wars," see Collins (1995), Daly (1990, 1991), Haraway (1991), Labinger (1995 and replies), Latour (1990), Oldroy (1990), Restivo (1995), Turnbull (1991), and also Feyeranbed (1993) and Laudan (1990).
93. Latour (1987), Pinch (1992).
94. Bourdieu (1975: 39).
95. Chubin and Restivo (1983).
96. Goonatilake (1993), Moghadam (1991), Sardar (1988), Shrum and Shevan (1995).
97. Benchimol (1990, 1995), Benchimol and Teixeira (1993), Botelho (1990), Cueto (1989), Fernandes (1987), Ferri and Motoyama (1979), Lopes (1964), Marques (1989), Morel (1979), Motoyama (1988), Niskier (1970), Paty (1992), Polanco (1985), Sagasti (1971), Schwartzman (1978, 1979, 1981, 1982), Stepan (1976), Valle and Silva (1981), Vessuri (1987).
98. Bernal (1972), Clark (1985), Cooper (1973), Dedijer (1968), Gaillard (1991), Navarro (1981), Polanyi (1968).
99. Morazé (1979), Morazé et al. (1980), Pannier (1979), Salomon et al. (1994), Silva (1980).
100. Breihl (1981), Laurell (1976, 1987).
101. See Almeida Filho (1989a, 1989b, 1990a, 1990b, 1991a), Costa (1990, 1993), Minayo (1995), Physis (1993), Possas (1989), Sabroza et al. (1995).
102. According to epidemiologist Paulo Chagastelles Sabroza (personal communication, 1992), the Brazilian and Venezuelan schools of epidemiology are more moderate than the radical historical materialism schools of Mexico and Ecuador. "Modera-

tion" accounts for the inclusion of the individual in the determination of the disease, as well as for the inclusion of structuralist and Foucauldian notions, which are usually rejected by most orthodox Marxist analysts.

103. See, for example, Minayo (1995).

104. Löwy (1992: 371), Mulkay (1972).

105. Horton and Aggleton (1989) argue for crossover between social and medical disciplines as a way to overcome the dead ends that develop in the study of AIDS.

106. See, for example, Amin (1989: 136).

107. Kuhn (1962). For an analysis of Kuhn's ambivalence, see Bourdieu (1975: 38).

108. See, for example, Mann et al. (1992), National Research Council (1989).

109. See, for example, Baldo and Cabral (1991), Packard and Epstein (1991), Sabroza (personal communication).

2. Politics and the Construction of Knowledge

1. See Altman (1987), Bolton (1989), Burkett (1995), Carter and Watney (1989), Crimp (1987, 1990), Daniel and Parker (1993), Feldman (1994), Gomez (1992), Kramer (1989), Mann et al. (1992), Nussbaum (1990), O'Neill (1990), Panem (1988), Pierret (1992), Pollak (1992), Scheper-Hughes (1994), Shilts (1987), Sontag (1989), Watney (1994), WHO (1985b, 1993c, 1994a).

2. See Risse (1988), Rosenberg (1989).

3. Gilman (1988).

4. Slack (1985).

5. Morris (1976), Rosenberg (1962).

6. Brandt (1987).

7. Crosby (1976).

8. Carvalho (1987).

9. Dubos (1965), Garrett (1994), McNeil (1976), Sigerist (1943).

10. See Von Giziky (1987).

11. Boston Women's Health Book Collective (1973), Ehrenreich (1973, 1978).

12. Nussbaum (1990: 108).

13. CDC (1981a, 1981b).

14. Shilts (1987: 66–67).

15. Laurie Garrett notes that even though Dr. Friedman-Kien took credit for reporting Kaposi's sarcoma, this disorder was first identified within the context of AIDS at the School of Medicine of the University of California at San Francisco.

16. Panem (1988: 160); see Gottlieb et al. *MMWR* 30: 250–252, Friedman-Kien et al. *MMWR* 30: 305–308, Gottlieb et al. *N Engl J Med.* 305: 1423–1431, Siegel et al. *N Engl J Med.* 306: 1439–1444, Masur et al. *N Engl J Med.* 305: 1431–1438.

17. See Shilts (1987: 68–69).

18. See Oppenheimer (1988).

19. See Sontag (1979, 1989).

20. See Gorman (1986), Mondragón et al. (1991).

21. This observation, made by Brazilian analysts that I interviewed at the beginning of my fieldwork, was actually quite ingrained in the population, as I could confirm a few years later (1993) in a presentation on AIDS prevention that was made in Casa do Brasil, a community association for Brazilian migrants in Lisbon, Portugal. The fact that the speakers stressed the increased risk of migrant populations was taken as an

insult by one member of the audience, who, as a "migrant," felt stigmatized by the statement, as if it implied that he was also promiscuous and gay.

22. For an accurate report of the history of media's approach to AIDS, see Kinsella (1989).

23. See Horton (1993).

24. See, for example, Davis and Whitten (1987), Caplan (1987), Fry (1995), Greenberg (1988), Guimarães (1977), Lancaster (1994), MacRae (1990), Mass (1990a, 1990b), Murray (1995), Parker (1987, 1989b), Parker and Carballo (1991), Plummer (1981), Whiten (1979).

25. Spontaneous comments from Latin Americans and Mediterraneans on the characteristics of the gay movement and gay culture indicated that they linked it to the Protestant ethic and the puritan tradition of controlling natural impulses, which created a transparency between the social and the intimate persona. While some of my Brazilian informants raved about U.S. gay culture, many others despised the "ghettoization" of homosexuality and what they saw as the regulatory and labeling manias of Americans. Instead, they praised libertarianism, which would allow impulses to be expressed independently of identities and free of regulation or disclosure. For a culturalist analysis of this discourse, see Parker (1991). The part about the Protestant ethic should be compared with the Roman Catholic (Iberic version) background of Brazilian culture, its unreachable ethical requirements, and its assumption of deviance and forgivable sin.

26. Shilts (1987: 117, 511).

27. In cities such as San Francisco and New York, an industry of identifiably gay services flourished. They included gay-owned businesses with gay employees catering to gay clients on either gay or general issues. They included bookstores, travel agencies, movers, real estate agents, restaurants, bars, and bathhouses, and came from a broad range of industries, including entertainment and commerce. They also included several professional support services, such as psychotherapy, medical referrals, dentistry, legal services, housing, twelve-step programs, and religious services.

28. The most talked about "new diseases" that followed the sexual revolution and the multiple-partner sexual pattern included hepatitis B, syphilis, venereal warts, and, by the mid-1970s, amebiasis, giardiasis, and shigellosis (Gorman [1986: 162]). As far back as 1978, "gay bowel syndrome" was reported (see Mass [1990a: 133]), even before HIV had been discovered. A single-topic medical book with the title *AIDS and Infectious Diseases of Homosexual Men* was published in 1984 by two doctors in New York (see Ma and Armstrong [1984]).

29. See Mass (1990a, 1990b).

30. See, for example, Burr (1995).

31. Halperin (1995).

32. In the Brazilian gay settings I observed, the "gay gene" theory was mocked as homophobic. Even gay men who were comfortable with their sexuality informed me that they did not like to attribute their sexual orientation (which they would rather call their "sexual option") to biological determinism. "If they tell me it's in the genes, I'll be heterosexual just to show that they are wrong," I heard once as a comment on the theory.

33. Kinsey et al. (1948, 1953).

34. See Altman (1987, 1988, 1995).

35. See Davis and Whitten (1987).

36. See, for example, Cagnon (1988), Carballo (1988), Carballo et al. (1989), Gorman (1986: 170).

37. See, for example, Cagnon (1988), Carrier (1989), Davies (1989), Davies and Coxon (1990), Herdt et al. (1991), Parker (1987, 1988), Parker, Herdt, and Carballo (1991).

38. The first time I read about this topic in print was in 1987–88 in the AAA newsletter. An article described Richard Parker's views on Brazilian sexual culture. Parker later published a number of articles addressing the issue, both from a theoretical and from an applied perspective (see, for example, Parker [1987, 1988, 1992]).

39. See Parker (1987, 1988, 1991a, 1992).

40. Loyola (1994), Paiva (1992), Parker (1992, 1994b), Parker, Mota, et al. (1994).

41. The popularization of the term "men who have sex with men," or MSM, did not necessarily meet with general success. A young person informed Michael Pollak (1988) that he considered it a useless euphemism, which sounded as though he planned on "going back to the closet" after having gone through so much hardship to acknowledge his sexual identity.

42. WHO (1988b, 1989a).

43. See, for example, Larvie (1995), Mota (1994), Valle (1994).

44. See Shilts (1987).

45. Shilts (1987).

46. See Fujimura and Chou (1995).

47. Lauritsen (1990).

48. *Spheric* (1994).

49. Duesberg (1987, 1989, 1992, 1994).

50. See Prescott (1994).

51. See, for example, Root-Bernstein (1993).

52. See Prescott (1994). To make his dismissal of the HIV hypothesis more convincing, Willner punctured his fingers with blood obtained from people who were HIV-positive (according to ELISA tests); this was done in public shows. In his view, the ELISA (enzyme-linked immunosorbent assay) test did not mean much because it gave a positive result for most infectious diseases. An odd spectacle of diversity in perspectives on AIDS could be seen at the Gay Community Center in New York on the evening of November 16, 1994. Upstairs, the group HEAL had invited Willner for a talk, while in the main room the most important gathering of AIDS activists evaluated the situation of the HIV epidemic in New York, the Town Meeting on the topic "Are We Surviving? The State of HIV Prevention in the Gay and Lesbian Community."

53. See Shilts (1987).

54. Kramer was charged with homophobia and anti-eroticism in a letter signed by Robert Chesley in the *New York Native,* December 21, 1981 (see Kramer [1989: 10] and Nussbaum [1991: 94]). Callen, who co-signed with Richard Dworkin and Richard Berkowitz the piece "We Know Who We Are: Two Men Declare War on Promiscuity" (*New York Native,* November 8–12, 1982), was attacked as a "sexual Carry Nation" and rebutted in the same newspaper through letters from infuriated readers (Nussbaum [1990: 98–99], Shilts [1987: 209–210]). A detailed description of the facts is given by an older, more reflective Kramer in his last book, *We Must Love One Another or Die* (Kramer [1998]), published after this manuscript was completed.

55. The book was severely criticized, but not just for presenting an image of

irresponsibility in the gay community. In fact, this criticism was later retracted by its authors, who admitted that if the community had listened to the warnings and acted more responsibly, many lives might have been spared (see, for example, Feinberg [1994]). In addition, *And the Band Played On* was heavily condemned for wrongfully portraying real people, alive and dead, as having thoughts, words, and intentions that the author could not possibly have known. Making a culprit of Gaetan Dougas, the Canadian air steward portrayed as "patient zero," and demonizing the promiscuous anti-hero, Shilts misrepresented a multi-centered epidemic and fed into popular myths and stereotypes.

56. Arno and Feiden (1992), Burkett (1995), Callen (1990), Kramer (1989), Mass (1990a), Nussbaum (1990), Shilts (1987), Van Vugt (1994).

57. Led by Martin Delaney, Project Inform wanted to shorten the long wait for lifesaving information and to systematize the knowledge of what was already used in the community based on empiric experiences. Commentators attribute Delaney's commitment to make drugs available, even if they are toxic, to his previous experience with trials for hepatitis, which he survived by using an experimental medication that produced neuropathy as a side effect; others, who did not have access to the medicine because of its toxicity, lost their lives. Project Inform's goal was "to conduct community research and monitor progress of patients using unapproved medications." This mission proved to be ahead of its time (see Arno and Feiden [1992: 32]). By then, it was not possible to fund such a venture. Their activities then shifted to making available information about experimental drugs and drugs that were available only in Mexico, where pharmacists sold rivabirin and isoprinosin without a prescription.

58. Perrow and Guillén (1990: 114).

59. See Altman (1995), Carter and Watney (1989), Feldman (1994), Watney (1994).

60. The equivalent to gay life in the seventies is depicted by Frederick Whiten (1995 [1979]) in the culture of *entendidos* (literally, "those who understand it," meaning "insiders"). The topic was also developed by Carmen Guimarães in an academic thesis (1977). According to a local myth (MacRae [1990]), it was the visit of the editor of *Gay Sunshine,* which produced the 1979 piece, that motivated the Paulistans to organize some sort of gay group, from which SomoS would develop. See also Mott (1995).

61. For a short but highly visible period of time (1978 to 1981), and gathering together accredited journalists, intellectuals, activists with political ambitions, writers, and social scientists from Rio de Janeiro and São Paulo, *O Lampião da Esquina* served as a forum for discussing the contradictory ideas and choices regarding homosexuality and the politics of sexuality.

62. SomoS was a São Paulo–based organization with a symbiotic relationship with the newsletter *Lampião.* These two social forces represent the peak in gay visibility and organizing, yet they succumbed before turning into something more substantial. By the time AIDS became an issue in Brazil, they no longer existed as organizations. However, they left personal and political marks on those who later became involved in the epidemic one way or another.

63. An exception is the Grupo Gay da Bahia (GGB) in the capital of the northeastern state of Bahia, which early on alerted its clientele of the "gay cancer" that had been reported in the United States, and whose leader, Mott, also preached mo-

nogamy. In Rio, the gay group Atobá, working out of a poor suburb with a mostly poor and working-class clientele, also included AIDS awareness in its activities. When internationally sponsored AIDS–NGOs such as ABIA (see Chapter 4) took the leadership of AIDS action, the small and poorly funded gay groups could hardly compete on the scene.

64. See Larvie (1995).

65. Mass (1990a: 143).

66. *The Native* was one of the first publications to give regular attention to the topic of AIDS, including it in its medical issues section, which was signed by gay physician Lawrence Mass. With time, most of its editorials came to challenge medical consensus about HIV or AIDS. In fact, editor Charles Ortleb became known for his vehement attacks on current mainstream beliefs, always writing "AIDS" within quotation marks.

67. The original founders were Larry Kramer, Paul Popham, Nathan Fai, Paul Rapoport, Lawrence Mass, and Edmund White (Mass [1990a: 122]).

68. In its second newsletter (January 1982), the GMHC warned the community that "Physicians are currently advising their gay patients, especially those who live in urban centers with large gay communities, to limit their sexual activity by having fewer partners and by selecting partners who are known to be in good health and who are themselves limiting the number of different partners with whom they have sex . . . apart from abstinence, monogamy represents the lowest risk potential" (Mass [1990a: 116]). This message traveled rapidly, and its wording was soon replaced by less alienating suggestions from the community, such as the practice of safer sex. Reality changed fast; a few years later, the use of those terms by epidemiologists and public health officers was radically rejected by activists. When an ad promoted by the Brazilian government in 1991 highlighted the phrase "reduce the number of sexual partners," it was severely criticized by activists and social commentators. The point was not reduce or fear sex, but to practice safer sex.

69. The lengthiest insider's account of the early years of the agency is given by cofounder Larry Kramer in his last book, which was published after this manuscript was completed.

70. Reinfeld (1994).

71. See Reinfeld (1994).

72. The boycott of Helms's funding sources included the most popular cigarette, Marlboro, whose ads had targeted and attracted modern gay men.

73. Shilts (1987: 557).

74. See Harrington (1994: 160).

75. Michael Callen was widely known for several reasons. A quite vocal PWA, he had a strong connection with his physician, Dr. Sonnabend, and supported the doctor's multifactorial hypothesis for AIDS. For some time, they believed AIDS was caused by a number of interacting variables, including *Cytomegalovirus*, the Epstein-Barr virus, repeated exposure to drugs and sperm, and the use of drugs, such as poppers. Evidence came from the fact that people who got sick with AIDS were also those with a higher number of sexual partners, and the correlate variables of more cases of sexually transmitted disease, more antibiotic treatments, more exposure to drugs, and a greater number of infections by other viruses and microbes. Speaking out against the promiscuity that he had once epitomized, Callen was bashed by the community, and several letters to the editor of the *New York Native* were sent and

articles written to express the disapproval of community members. He outlived his diagnosis for many years, and his faith in the multifactorial hypothesis made him distrust antiretrovirals such as AZT. For that, he continued to be a polemic figure.

76. Rather than being a politico, Campbell was a festive member of the "sisters of perpetual indulgence." His was one of the earliest public disclosures as a person with AIDS, and he was sometimes referred to as the "Kaposi's sarcoma poster boy."

77. PWA coalition (1985).

78. Contradictory narratives depict that transition as either smooth or almost nonexistent. Even though the current figures furnished by the agency portray a quite diverse array of clients, a number of people from a variety of ethnic groups reported feeling excluded by an organization that had a predominantly white gay history and culture.

79. Harrington (1994), Nussbaum (1990), Wolfe (1994).

80. Boston Women's Health Book Collective (1973).

81. The most well-known of the new gurus was Louise Hayes, whose "Healing Circles" extended from California to the East Coast. For an account of the New Age movement, see Ross (1991).

82. AL 721 was a lipid concentrate obtained from egg yolk. It was assumed that the lipids might prevent the virus from binding to the cell membrane. According to analyst Bruce Nussbaum, who takes a close look at the "business" of drug development, AL 721 was not developed because of the rapid development of AZT, which he also attributes to a desire for profits rather than for achieving a scientific and therapeutic breakthrough (Nussbaum [1990]).

83. See Nussbaum (1990: 251).

84. Gay physician Lawrence Mass, who, as a *Native* writer and founder of GMHC, had been involved with the epidemic since it began, was able to evaluate the relationship between the gay community and the medical establishment from both sides. Distancing himself from patronizing and insensitive medical authorities as well as from radical, anti-science activists, he synthesized his "centrist" position as "it's possible to be aware and critical of pseudo-science and homophobia throughout the history of medicine and science, while at the same time respecting that, whatever the political obstacles, our understanding and control of the epidemic must eventually come from within our mainstream medicine and science" (Mass [1990a: 123]).

85. Nussbaum (1990: 221).

86. See Haraway (1992), Martin (1994).

87. The development of the potentially anti-HIV new drug started in the NIH laboratories (see Nussbaum [1990]). After LAV/HTLV–III/HIV was isolated, NIH researchers screened compounds for antiretroviral activity. Suramin, the first candidate, was discarded after negative results. Compound S didn't have a patent (or any apparent use), but showed in vitro antiretroviral activity. It turned out to be azydothymidine (AZT, now more commonly called zidovudine), and it would become the first magic bullet for AIDS. To develop the drug, NIH researcher Sam Broder negotiated with the pharmaceutical corporation Burroughs-Wellcome for sponsorship and a supply of the drug for clinical trials.

88. See Nussbaum (1990).

89. To avoid taking a placebo instead of the "only hope," some patients who were enrolled in these trials shared their assigned pills with each other to increase their chances of taking the real drug.

90. AZT was approved on March 17, 1987, and reigned as the sole antiretroviral on the market for five years. A similar drug (i.e., nucleoside analogue), ddI, was approved on October 9, 1991 (Arno and Feiden [1992: 18]).

91. Arno and Feiden (1992: 6, 55).

92. AZT was for some time the most expensive medicine ever. Other medications that were developed later to treat and prevent AIDS-related infections were even more expensive. Such was the case with foscarnet, a drug produced by Astra laboratories to treat CMV-related retinitis and blindness in individuals who did not respond to Syntex's ganciclovir. With a cost of $25,000 a year, foscarnet turned Astra into a major target for AIDS activists (see Harrington [1992: 6]). At the Eighth and Ninth International Conferences on AIDS, held in Amsterdam and Berlin, their anger about the corporation peaked. Everyone in attendance was given leaflets depicting Astra's "Astranomic greed," and the company's exhibit booth was destroyed.

93. Arno and Feiden (1992: 118).

94. Nussbaum (1990: 233).

95. Arno and Feiden (1992), Black (1986), Nussbaum (1990), Harrington (1994).

96. Nussbaum (1990: 74–75).

97. Harrington (1994: 161–162).

98. Sonnabend's popularity was higher among PWAs, social commentators, and journalists (see, for example, Black [1986], Burkett [1995], Nussbaum [1990]) than among his fellow scientists. Scientists were, for the most part, quite comfortable with the germ-theory approach, which targeted HIV as the only cause of AIDS, and pursued research to find a single "magic bullet" to destroy it (see Goodman [1995]). The few scientists who challenged this approach—such as Peter Duesberg and Robert Root-Bernstein—were considered mavericks, and their ideas were rejected by the mainstream scientific community (see Duesberg [1987, 1989, 1992, 1994], Root-Bernstein [1993], Coulter [1987], Lauritsen [1990], and for a comprehensive analysis, see Fujimura and Chou [1994] and, particularly, Epstein [1996]). Sonnabend, like most PWAs, did not oppose the medical establishment; nor did he fully dismiss the role of HIV in AIDS, as Duesberg and others did. The contrast between this community physician and those who accepted mainstream biomedical wisdom is more a matter of background and focus than of incompatible theories, as was the case for Duesberg and his followers.

99. Black (1986), Moraes and Carrara (1985a), Mass (1990a), Nussbaum (1990), Sonnabend et al. (1984).

100. AmFAR grew out of the merging of the AMF and a parallel organization in California, which was headed by Dr. Michael Gottlieb and promoted by Hollywood legend Elizabeth Taylor. Gottlieb had been Rock Hudson's doctor, and Rock Hudson had allocated funds in his will for AIDS research. AmFAR later became the major source of funds for New York's CRI (Arno and Feiden [1992: 50], Nussbaum [1990: 225]).

101. In a polemic cover of the *Village Voice,* Callen was presented as someone who saw love, a good physician, and Classic Coke (not dismissing luck) as the keys to his survival.

102. Callen (1990).

103. Harrington (1994: 174–176).

104. New York/San Francisco differences emerged clearly over this topic. Martin Delaney and Project Inform supported the trial, while Callen and, officially, New

York's CRI opposed it because they believed the drug was too toxic. Yet underground trials were executed, some of them even involving CRI staff members. The drug turned out to be too toxic, and the trials were stopped after some people died.

105. See Harrington (1994: 171–172).

106. See Elbaz (1990, 1991, 1992), Gamson (1991), Wolfe (1994).

107. Kramer (1978).

108. See Shilts (1987).

109. See Mass (1990a: 120–121).

110. Kramer (1985).

111. Arno and Feiden (1992: 8).

112. Nussbaum (1990: 328).

113. See Kramer (1989).

114. Kramer (1989: 134).

115. Wolfe (1994).

116. Wolfe (1994).

117. Wolfe (1994: 218).

118. See Wolfe (1994: 219).

119. Wolfe (1994: 224).

120. See Kramer (1989), Wojnarovic (1991).

121. Kramer (1989).

122. See Crimp (1990), Wolfe (1994).

123. For a thorough documentation, see Crimp (1990).

124. Different from the older "gay clone" from Christopher Street—with his trimmed mustache, neat hairdo, lumberjack flannels, and jeans jacket—the "new clones" shaved their heads, moved the pocket bandanna to their heads, inverted their baseball caps, pierced both ears, and moved to the East Village and Chelsea rather than the West Village or the suburbs. The new fashion statements, themselves influenced by African-American hip-hop artists and enthusiasts, were rapidly adopted by other populations as well.

125. Elbaz (1992), Signorile (1993).

126. For a full account of this process, see Elbaz (1992) and Wolfe (1994: 225), as well as Feinberg (1995) for an experiential narrative. The compilation of data by ACT UP's special committees represented a unique form of assembling and synthesizing the state-of-the-art medical knowledge as well as unofficial knowledge. The *Treatment and Research Agenda* and the *Countdown: 18 Months,* compiled by the Treatment and Data Committee, that I brought to my field site in Brazil, proved to be a priceless tool for research and was praised by activists and doctors alike.

127. Lack of representation by a variety of social groups was one of the most cited criticisms of ACT UP. However, this topic was frequently addressed in general meetings, where the membership was often split between giving the priority to AIDS, regardless of the social implications associated with the struggle, or making sure that every intervention addressed every political aspect of the AIDS crisis.

128. Nussbaum (1990: 323).

129. See, for example, Crimp (1990), Elbaz (1990, 1991, 1992), Feinberg (1994), Gamson (1991), Wolfe (1994).

130. ACT UP (1990).

131. See Wolfe (1994: 231–237).

132. Harrington (1992: 1).

133. This was an observation directed to TAG members by a journalist during a press conference held at the Ninth International Conference on AIDS, in Berlin, 1993.

134. *Concorde* refers to a French–British multi-center study whose results were released at the Ninth International Conference on AIDS, which was held in 1993 in Berlin. For many physicians whose "faith" in AZT was doubled by evidence of its ability to reduce patients' symptoms, the data felt like a bomb. The doctors had been administering AZT to asymptomatic HIV-infected people in the honest belief that it delayed the onset of symptoms, which was medical dogma until that moment.

135. Harrington (1992).

136. According to Andrew Sullivan (1996), even the most skeptical of TAG's members accepted the evidence given by the first results of using the triple cocktail with protease inhibitors: the lowering of HIV load into "undetectable levels" in people with AIDS. Caution about the virus "hiding" somewhere else or about creating multi-resistant variants of the virus gave way to acceptance of the good news, possibly because it was released during a period marked by activist burnout.

137. This was named after an unorthodox scientist who dared challenge dogmas; see Keller (1983).

138. Political funerals were very special forms of public intervention; the body of someone who had just died of AIDS was carried during a political demonstration to allow friends, families, and supporters to mourn the death and to promote public awareness of the tragedy behind AIDS. During the demonstration in front of the White House, someone's urn was carried and the ashes were thrown across the gates.

139. GMHC (1995: 1–2).

140. Sullivan (1996).

141. See, for example, Burkett (1995), Lapierre (1990).

3. Sponsoring Global Action

1. See Mann et al. (1992: 231).

2. More precisely, the United States had 339,250 reported cases of AIDS in 1993, which is about 40 percent of the 851,628 cases reported from around the world at the time (source: *Weekly Epidemiological Record*, January 14, 1994). It should be noted that a difference in reporting patterns is a biasing element and may distort the actual distribution of the epidemic.

3. Fleming et al. (1988), Mann et al. (1988), Mann et al. (1992), Panos (1989, 1990), Sabatier (1988), WHO (1985a, 1985b, 1988b, 1988c, 1989a, 1989b, 1989c, 1992, 1993c, 1994b).

4. See Dazinger (1994), Dhillon and Philip (1994), Heise and Elias (1995), Panos (1990), Rothermel (1988), Weill et al. (1994), WHO (1994b).

5. WHO (1993c: 99).

6. WHO (1992: 7, 1994b: 9).

7. Mann et al. (1992: 107).

8. As a freelance reporter at the Eighth International Conference on AIDS, which was held in Amsterdam in 1992, I observed a not very friendly exchange of words between Michael Merson, then coordinator of the Global Programme on AIDS, and Jonathan Mann, its former coordinator, who was at the time head of the Global AIDS Policy Coalition. The GPA's leader seemed appalled and surprised by what he

considered an exaggerated projection of cases by the Global Coalition; this agency had chosen the Delphi method (Mann et al. [1992: 106–107]) to estimate future occurrences of the disease around the world.

9. WHO (1994b: 17).

10. WHO (1994b: 17).

11. WHO (1993c: 3).

12. WHO (1992: 7).

13. The Global AIDS Policies Coalition is a not-for-profit partnership formed by persons involved in the fight against AIDS at a global level. It was created and led by Jonathan Mann after he left the position of coordinator for WHO's Global Programme on AIDS in 1991.

14. Mann et al. (1992: 19).

15. Reflecting a mean between the year a national leader first addressed the issue of AIDS and the year of the creation of a national program, the dates were 1982 for North America; 1984 for Oceania; 1986 for Western Europe, Latin America, Sub-Saharan Africa, and North and East Asia; 1987 for the Caribbean and Southeast Asia; and 1989 for Eastern Europe (Mann et al. [1992: 20–21]).

16. Mann et al. (1992: 20–21).

17. Mann et al. (1992: 30).

18. Mann et al. (1992: 27).

19. Panos (1989, 1990), Sabatier (1988).

20. Mann et al. (1992).

21. WHO (1994b).

22. I synthesized these points in a few media articles after the VIII International Conference on AIDS, held in Amsterdam in 1992 (e.g., SIDA—o Desafio Global, *Expresso,* August 1992).

23. Evidence for this statement was obtained from observation of the AIDS international movement during a significant period of time (1988–1995). During this time period, I obtained information from alternative conferences held during the Sixth International Conference on AIDS in San Francisco (1990), and from attendance as an observer and journalist at the Seventh, Eighth, and Ninth Conferences on AIDS, held, respectively, in Florence, Amsterdam, and Berlin. More information was gained through personal interviews with AIDS activists and AIDS program officers from North and South America, Europe, and Africa; members of international AIDS networks; and WHO officers. These interviews focused on the international AIDS movement and AIDS activism in New York and Rio de Janeiro.

24. Harrington (1992).

25. See As Sy (1995), Cash et al. (1994), Filgueiras (1994), Kit and SAfAIDS (1995), Mussauer (1994), Rowe (1994).

26. WHO (1985a: 98).

27. WHO (1985a: 99).

28. The conference was eventually held annually and included among its meeting sites Paris, France (1986), Washington, D.C. (1987), Stockholm, Sweden (1988), Montreal, Canada (1989), San Francisco, California (1990), and Florence, Italy (1991). The 1992 conference was scheduled for Cambridge, Massachusetts, but had to be relocated because the organizers were opposed to U.S. restrictions on entry into the country of people with AIDS; it consequently was transferred to Amsterdam, Holland. In 1993 the conference took place in Berlin, Germany; in 1994 it took place

in Yokohama, Japan, which became the first Asian host site. Since AIDS research had become part of "normal science" by that time (owing, in part, to the fact that researchers were coming up with more predictable results), the number of groundbreaking findings was no longer considered worth the effort of putting together these expensive conferences every year. Thus, they became biennial; Vancouver hosted the 1996 conference. Also, smaller and more specialized meetings became the preferred means of sharing information; to reach a greater number of people, conference organizers could take advantage of the latest in communications technology.

29. Shilts (1987: 261).

30. The ideas overlapped and interacted; a conceptual middle ground had to be reached. The definition of two epidemic patterns, which will be described later, partly resolves the opposition.

31. My introduction to global data on AIDS was in 1988 through the non-trade sheets published by the Panos Institute—later published as the Panos Dossier (Panos [1989])—and available at the New York Public Library. While reviewing some of this material, I noticed that the distribution of AIDS reported was very uneven across the African continent. This made me think that social variables—such as economic development, surveillance mechanisms, and perhaps ties to international agencies or to other governments—representing the aftermath of European colonialism might be influencing the figures. For example, the former Belgian colonies of Rwanda, Burundi, and Zaire had the highest incidences of AIDS cases. I learned later that, coincidentally, some of the researchers who were most heavily involved with AIDS were from the Belgian Tropical Medicine Institute and had developed lines of cooperation and assistance with their country's former colonies; see Shilts (1987: 392) and Garrett (1994). Also, former Portuguese colonies, such as Angola, Mozambique, Cape Verde, and Guinea Bissau—which are scattered across the continent—reported a significant number of cases of HIV-2 at a time when this virus seemed to be confined to West Africa. I learned later that HIV-2 had just been isolated for the first time in Lisbon, Portugal, with technical support from the Pasteur Institute HIV team. Researcher Odette Ferreira pursued the analysis of blood samples of a number of AIDS patients from the tropical disease wards of Egas Moniz hospital who had repeatedly tested negative for HIV with the regular tests. The influence of other variables—such as war, traffic, and migration—became increasingly clear as social researchers contributed to the study of AIDS in Africa.

32. WHO (1985a: 98).

33. Baldo and Cabral (1991), Garrett (1994), Mann et al. (1992), Packard and Epstein (1991), Preston (1994), Schoepf (1991).

34. See, for example, Gould (1993).

35. Mann et al. (1992).

36. Mann et al. (1988: 84).

37. The popular/media version of this association drew not only on stereotypes of reckless black sexuality but also on several other "exoticisms" (from circumcision to clitoridectomy to scarification) as possible means of HIV transmission in African AIDS.

38. Daniel (1985, 1989), Daniel and Parker (1990), Guimarães et al. (1988), Ramos (1990, 1992).

39. See ABIA (1988b, 1988c), Daniel (1985, 1989), Guimarães et al. (1988), Ramos (1990).

40. Guimarães et al. (1988).

41. E.g., one of the odd risk categories used in the early epidemiologic forms in Brazil was "travels to Haiti."

42. See Guimarães et al. (1988).

43. Parker (1987, 1988, 1991a).

44. That critique was not universally shared. Self-identified gay groups, even though small in number and clientele, accepted the U.S. model. So did the government and the epidemiologists who jumped on the AIDS research bandwagon. Most medical doctors, while recognizing the possibility that a number of characteristics of Brazilian sexuality might invalidate that model, did not challenge the mainstream model, which was the one they had been trained to accept during their medical training. In 1991, the media speculated grandly about the possible "Africanization" of Brazilian AIDS, meaning that predominantly homosexual transmission was being replaced by predominantly heterosexual transmission.

45. Carrier (1995), Lancaster (1994), Murray (1995).

46. See, for example, Parker and Carballo (1990, 1991), Parker, Herdt, and Carballo (1991).

47. Mann et al. (1992). The authors depart from the original three patterns of epidemiology developed by some of them while at WHO (Mann et al. [1988]) and update the global epidemiology of AIDS by creating ten categories, which they refer to as "Geographic Areas of Affinity."

48. See, for example, Albrow and King (1990), Bergensen (1990), Featherstone (1990), Featherstone, Lash, and Robertson (1995), Friedman (1994), Hannerz (1992), Ianni (1992), King (1991), Robertson (1992).

49. See, for example, Barnett and Cavanagh (1994).

50. WHO (1962: 30).

51. WHO (1987).

52. See National Health . . . (1993), WHO (1962, 1975, 1976, 1988a).

53. Created in 1907 in Paris, OIHP was the first agency to address the problems of international health. WHO, which was created following the Refugees of War Commission, started its activities in 1948.

54. WHO (1976: 18).

55. Peabody (1995).

56. Garrett (1992: 825).

57. Garrett (1992, 1994).

58. Torino and Creese (1990), WHO (1979, 1980, 1981, 1986a, 1993a, 1993b).

59. WHO (1994a).

60. Peabody (1995: 735).

61. WHO (1994a: 15).

62. WHO (1979: 40).

63. WHO (1979: 41).

64. WHO (1979: 42).

65. WHO (1979: 43).

66. WHO (1979: 44).

67. WHO (1981: 86).

68. WHO (1981: 87).

69. WHO Chronicle 37 (1983: 4).

70. WHO (1987).

71. WHO (1994a: 15).
72. WHO (1986b: 52).
73. WHO (1985b: 211).
74. WHO (1985b: 211).
75. Those are now known as hepatitis C and Delta.
76. WHO (1988b: 2).
77. WHO (1988c: 1).
78. WHO (1992: 19, 1994b: 4).
79. WHO (1992: 19).
80. WHO (1992: 19, 1994a: 4).
81. WHO (1988b).
82. WHO (1985b: 211).
83. See Mann (1991).
84. See, for example, As Sy (1995), Cash et al. (1994), Filgueiras (1994), Kit and SAfAIDS (1995), Mussauer (1994), Rowe (1994).
85. Evidence in support of this statement was provided during field work in Brazil, where I observed the expansion of the Alternex network in 1991, years before the popularization of Internet. The first CD-ROM device I ever saw was being used at a Brazilian NGO at least as early as 1990. Fax machines were a priority, even for the smaller NGOs. All of these devices were used frequently.
86. See Garrett (1994: 459–465).
87. Cuba was the only example where traditional public health measures were implemented, as a result of the combination of unique characteristics: an exceptional national health service and an authoritarian regime. Reminiscent of the one Pacific island that was able to escape the 1918–1919 influenza pandemic by not allowing anyone in, Cuba was able to keep the AIDS epidemic under control and at levels that were unattainable in any other country. Traditional public health measures were implemented, including widespread testing, contact tracing, and quarantine. The cost of these measures in terms of individual human rights was not a main concern for the Cuban regime, whose priorities were in the public sphere. This approach was not transferable to any other political setting, and the Cuban model was condemned by human rights activists throughout the world. Oblivious of the individual rights and self-centered narratives, including an impressive account by Cuban exile Reynaldo Arenas (1993), Cuban medical doctors focused their action on providing medication, including interferon, as well as good-quality housing for the lifelong quarantine (see Perez et al. [1994]).
88. See, for example, WHO (1988c, 1989a, 1989b).
89. See, for example, WHO (1988c).
90. Shilts (1987: 156).
91. Panos (1989: 170).
92. WHO (1988c: 6).
93. There are many examples to illustrate this point. Even non-Catholic New York endured the pressure of the Catholic Church over the issue of condoms. In Brazil, one of the early campaigns portrayed a woman holding a condom and suggesting *previna-se* (literally, "prevent yourself," or "beware"). The Catholic Church vetoed the poster, which was released without the condom—and with hardly any message.
94. WHO (1988c).
95. WHO (1989a).

96. Evidence in support of this statement comes from observations and informal conversations in 1991 with Brazilian NGO leaders as well as community-based physicians. These individuals did not, however, represent the totality of Brazilians, and the new coordinator, even though less flamboyant, proved to be engaged with the developing countries and Brazil in particular.

97. See Global AIDS . . . (1991).

98. This was a very special conference. Its location was chosen as an alternative to Boston/Cambridge because there were restrictions on the entry of HIV-positive visitors into the United States. That restriction conflicted with the spirit of WHO's AIDS policies.

99. Mann et al. (1992).

100. Mann et al. (1993).

101. See ABIA (1992, 1993a).

102. Filgueiras (1993, personal communication). Also, UNAIDS (1997).

103. Peabody (1995: 735).

104. Peabody (1995: 735, 741).

105. Peabody (1995: 736).

106. Peabody (1995: 741).

107. See Garret (1994: 476, 481).

4. Local Action

1. Panos (1989), Sabatier (1988).

2. See, for example, Chirimuuta and Chirimuuta (1989), Panos (1989), Sabatier (1988).

3. Daniel (1990: 11).

4. See, for example, Mann et al. (1992).

5. See Daniel (1985), Ramos (1990).

6. Ministério da Saúde (1987, 1988a, 1988b).

7. Brazilians sometimes argued that "sida" would be easily confused with "Cida," a common female nickname, short for Aparecida or Alcida. In their view, the choice of a personal name for a disease would stigmatize women with either name. They did not seem to know that "aids" as an English noun (and verb) pre-existed the epidemic.

8. See Ramos (1990: 9–10).

9. See Cohn (1995: 85–86).

10. A similar association was reported in Tanzania, where the Swahili word "hella," money, was used for the new disease, which affected rich traders and fishermen (WHO 1994a: 3), much in the same way it affected well-traveled gay men in Brazil. Those cases happened to be more visible, like the American cases had been more visible internationally.

11. Moraes and Carrara (1985a).

12. Moraes and Carrara (1985a: 17).

13. Granjeiro (1994).

14. See, for example, Estado de São Paulo, May 18, 1991.

15. O Globo, November 15, 1987.

16. Folha de São Paulo, August 8, 1991.

17. Folha de São Paulo, October 3, 1991.

18. The possibility of female-to-male HIV transmission raised intense debates in Brazil. On the basis of a small body of research, physicians from the AIDS unit of the hospital Gaffrée e Guinle stated publicly that there was no evidence of such transmission. Even though the claim was dismissed by a number of local doctors and activists as well as by the literature, it remained an ideological contribution to the already strong resistance regarding the practice of safer sex. As a result, men, oblivious of their own serostatus, avoided condoms when having sex with women, to the latter's blatant disadvantage in a context where gender empowerment is unbalanced to begin with. As a pragmatic choice, most prevention campaigns focused on the use of condoms as the only way to avoid HIV infection. Easier to manage than issues such as power, negotiation, gender, and the like, the focus on condoms had its own problems, particularly given their high cost: in 1991, one unit cost more than the equivalent of one U.S. dollar, while the minimum monthly wage was less than $100.

19. See Ramos (1990: 13).

20. Two contrasting examples from 1989 are rock singer Cazuza and TV-Globo actor Lauro Corona. The young, handsome Corona hid his condition and continued to act in a soap opera until he could no longer stand up. He died in denial, appalling a multitude of fans and friends. In his memory, other actors and actresses campaigned against the prejudice that surrounds AIDS, so that no one else would have to endure the same type of solitude. At about the same time, Cazuza (whose family ranked higher in Globo itself) came out as a PWA. His disclosure was used as a tool against the disease, which he fought with all possible strategies. He became the local AIDS poster boy, lending his face to the front pages of mainstream magazines and serving as a role model for disclosure and activism. His mother, Lúcia Araújo, also became involved in anti-AIDS activities; she created a foundation after his death (in 1991), the Sociedade Viva Cazuza, which provides support and health care to children with AIDS.

21. Isto É 6.15.85, quoted by Moraes and Carrara (1985b: 25).

22. The media's morbid fascination with AIDS included an eager search for "aidético" (the medical name initially given to AIDS patients in Brazil, later termed politically incorrect by PWA organizations), even in hospital settings. Once a doctor commented on how she had disappointed a journalist by stating that she handled AIDS as if it were any other disease. While health care professionals tried to make the treatment of people with AIDS part of the normal routine of medical care, the media insisted on its "abnormality." For example, the magazine that employed the "disappointed journalist" decided not to publish the story, because it projected the "normality" of the professional care of AIDS. For a lengthy analysis of AIDS treatment in Brazilian media, see Galvão (1992a).

23. Betinho (Herbert de Souza) headed the large umbrella NGO known as IBASE (the Brazilian Institute for Social and Economic Research), which hosts a number of applied social research projects. IBASE pioneered the management of large-scale documentation projects, electronic networking and international conferencing, public intervention projects on social causes, and critical dialogues with governments about specific issues. Betinho was often consulted by progressive ministers and presidents. A charismatic figure, in 1994 he led the campaign of "citizens against famine," which almost brought him the nomination for the Nobel Peace Prize (his candidacy was jeopardized by his past acceptance of illegal gambling money as emergency funding for the AIDS organization ABIA at a time when it was at the verge

of bankruptcy). Betinho lived productively and joyfully in spite of his declining health. He welcomed enthusiastically the triple cocktail, even though his body did not accept the treatment for too long. He died in 1997.

24. This statement is based on interviews and exchanges collected between the years 1989 and 1992, primarily in Rio de Janeiro. People who had participated years before in now extinct gay groups such as SomoS, or who were gay but had never participated in a gay group, argued that their participating with broader groups of affected people—including blood recipients and women—was a better strategy for fighting AIDS and spreading messages about prevention. Some members of the existing gay groups, such as the Grupo Gay da Bahia and Atobá, preferred to continue to fight AIDS through "gay-identified" activities. The GGB remained very critical of what became mainstream AIDS activism, particularly the activities of the large agency ABIA in Rio.

25. For many of my "informants," this was a very alien and very "American" notion. In their analysis of early media responses to the new epidemic, Moraes and Carrara (1985b) criticized the rapid adoption by the media of the term "gay community." These authors considered the literal translation of the term as culturally inadequate within the Brazilian context at the time. However, with the implementation of the fight against AIDS and the strengthening of community-based organizations, the term *comunidade homossexual* (gay community) became more frequently used. In 1995, Rio de Janeiro hosted the annual International Lesbian and Gay Association (ILGA) meeting, an event in which federal and state government officers praised the importance of the involvement of the gay community in the fight against AIDS.

26. Racism in Brazil deserves a separate analytical treatment that is beyond the scope of this book. The pervasive form of local racism is nevertheless relevant in several ways. Not only does it provide a paradigm to better understand the local sexual way, but it also has a subtle impact on topics that appear without reference to race, and where it is yet present, like the "pauperization" of the epidemic, or the "Africanization" of Brazilian AIDS. (For a wider discussion, see Bastide and Fernandes [1951], Burdick [1995], Fernandes [1979], Fry [1982, 1989], Skidmore [1992], Sheriff [1995], Twine [1995], and the later Twine [1997].)

27. Mott (1995).

28. See Guimarães (1977), Whiten (1979). *Entendidos* means literally "those who understand," suggesting that they have access to a hidden knowledge about sexuality—and to its practices as well.

29. MacRae (1990).

30. Fry (1982).

31. Parker (1987, 1991a, 1992), and also Fry and MacRae (1985), Perlongher (1987).

32. Guimarães (1977), Terto (1988).

33. Mott (1985, 1988a, 1988b).

34. Fry (1982), Parker (1991a).

35. See McRae (1990).

36. The environmental Earth Summit in Rio 1992 and the Conference on Women in Beijing 1995, for example, would each have their own separate NGO forums.

37. Altman (1995), Arno (1989), Nelkin et al. (1991), Pollak (1994), Van Vugt (1994), Watney (1994), Wolfe (1994).

38. The Instituto Superior de Estudos de Religião (Institute for the Study of Religion) is another umbrella NGO whose scope goes much beyond its title, and

which promotes social intervention programs as well as research. It is funded primarily by religious donor agencies from Northern European countries such as Holland, Sweden, and the United Kingdom.

39. ISER (1985).
40. Rede Brasileira (1989a: 4).
41. Valinotto (1990: 6-7).
42. Solano (1992: 8).
43. Ministério da Saúde (1994a: 91).
44. Ministério da Saúde (1994a: 92).
45. See Landim (1988: 24).
46. See Reily (1995).
47. Landim (1988: 25).
48. Fernandes (1988: 8).
49. Fernandes (1988: 9); see also Fernandes and Carneiro (1995: 72).
50. Fernandes (1988: 9), Fernandes and Carneiro (1995).
51. See Landim (1988: 31).
52. Fernandes and Carneiro (1995: 75).
53. Reily (1995: 22).
54. Solano (1992: 9).
55. See Landim (1988: 47–48).
56. See Solano (1992: 9–10).
57. Galvão (1992b: 7).
58. Solano (1992: 12).
59. WHO (1989c).
60. WHO (1989c).
61. ABIA (1989b: 12).
62. Galvão (1992b: 10).
63. Galvão (1992b: 10).
64. Rede Brasileira (1989a: 3).
65. Rede . . . (1989a: 5).
66. Mann (1990).
67. ABIA (1989d), Rede . . . (1989b).
68. Rede . . . (1989c).
69. Rede . . . (1990).
70. Galvão (1992b: 11).
71. Galvão (1992b: 11).
72. Government officers often commented that it was easier to work with this type of group than with the self-assigned AIDS–NGOs. I observed that the gay groups Atobá in Rio and GGB in Bahia were more willing to work with the board of health or with medical institutions than, for instance, ABIA, whose critical stance often prevented it from cooperating with government- or medical-sponsored pragmatic programs. Prostitute groups also cooperated with the government through such joint programs as the ISER-affiliated group Prostituição e Direitos Civis (prostitution and civil rights), which produced a number of leaflets (see, for example, ISER [1990]) under the program "Previna" (see Braiterman [1991a, 1991b]).
73. Valinotto (1990, 1991), Parker (1994a: 40).
74. Solano (1992: 19).
75. See AHRTAG (1990a).
76. ABIA (1990a, 1990b).

77. AHRTAG (1990b: 1, 11).

78. Bouchara (1991: 9).

79. Mann et al. (1992).

80. With the increasing use of *Pessoa vivendo com AIDS* (PWA), the term *Doente de AIDS* (AIDS patient) was abandoned.

81. See Parker (1994a: 42).

82. Guimarães et al. (1988).

83. ABIA (1988a).

84. ABIA (1988a).

85. See ABIA (1988a).

86. See GGB (n.d.).

87. ABIA (1991d), Solano (1993).

88. Fernandes (1994).

89. Monteiro et al. (1994).

90. See, for example, Parker and Galvão (1994), Parker, Bastos, et al. (1994).

91. Terto (1993), ABIA (1994), Quemmel (1994), Parker, Mota, and Rodrigues (1994).

92. See Gaspar (1992).

93. Lair Guerra de Macedo Rodrigues headed the first National Program on AIDS and STDs (1986–1990). She was replaced during Fernando Collor's presidency, only to be returned to her post when Collor replaced his corrupt minister of health, Alceni Guerra, with INCOR's surgeon Adib Jatene in 1992. Previously a biomedical sciences professor at the University of Brasília, Lair Macedo was personally connected to the international bureaucracies of PAHO and WHO, and her curriculum vitae includes time spent at the CDC in the United States. As a woman, a Northeasterner, and a Baptist, she challenged at every instance the usual profile of power (male, Southeasterner, Catholic or laïc) with an unusual amount of political savvy. Her career was tragically interrupted by a car accident in 1996.

94. See, for example, Ministério da Saúde (1987, 1988a, 1988b).

95. For a polemic, see ABIA (1988c: 2) and Mott (1988c: 6).

96. See ABIA (1988d, 1988e, 1988h, 1988i, 1988j), Ramos (1988: 6); see also Tema/Radis (1988).

97. See ABIA (1988c: 2, 1988f, 1989a).

98. ABIA (1989a), Guedes (1989).

99. See ABIA (1988f, 1989c); see also Guimarães et al. (1988: 4–5), Ramos (1989: 9), Parker (1989a: 10).

100. Fry (1982), Fry and McRae (1985), McRae (1990), Perlongher (1987, 1992), Parker (1987).

101. See, for example, Parker (1987, 1988, 1989b, 1991a, 1992).

102. ABIA (1988d, 1988e, 1988j), Ramos (1988: 6).

103. Landim (1988: 45).

104. ABIA (1991b, 1991c).

105. See Schwarzstein (1992: 3–4).

106. Ministério da Saúde (1994a).

107. The campaign that was launched in early 1994, Você precisa aprender a transar com a existência da AIDS, was welcome by NGOs and the public. It included sketches depicting young people addressing candidly and in attractive ways the issues of safe sex and needle use.

108. See, for example, Ministério da Saúde (1994b).

109. This observation was made by some of the international delegates to the conference. The view that "ABIA turned into a publishing house" instead of promoting the development of new AIDS therapies was a remark made by ex-coordinator Walter Almeida, M.D., to *Jornal do Brasil*, March 18, 1995, in an interview about the potential of passive immunotherapy (based on the infusion of the plasma of long-term asymptomatic HIV-positive people) in reducing the severity of AIDS.
110. See Schwarzstein (1993a), ABIA (1993b).
111. Reily (1995: 24).
112. Ministério da Saúde (1994a: 83).
113. The typical poverty-related parasitic and infectious diseases required brief treatment, for the patient either died quickly or was fully cured without sequelae. Doctors specializing in infectious disease often mentioned that they had never developed personal ties with their patients until they started treating patients with AIDS.
114. Daniel (1989).
115. Gurgel (1988), GAPA–RJ (1989).
116. Gaspar (1992: 2).
117. See Beloqui (1994), Grupo Pela VIDDA et al. (1992), ONGS/AIDS do Brasil (1991), Schwarzstein (1993b), Sutmoller et al. (1994).
118. ABIA (1991a, 1991b).
119. See ONGs/AIDS do Brasil (1991), Schwarzstein (1993b).
120. Daniel (1989).
121. See ONGs/AIDS (1991).
122. Grupo Pela VIDDA et al. (1992).
123. See Grupo Pela VIDDA et al. (1992).
124. See Marques (1993).
125. See Green (1995).
126. Ramos (1989).

5. Central Problems in a Peripheral Landscape

1. Brazilian elites have always traveled abroad in search of "culture" and "identity." In the nineteenth and the first half of the twentieth centuries, they usually went to France, which they then considered the center of the civilized world. After the Second World War, the United States became for them the model of modernity and progress. It is common today to see members of the Brazilian middle classes travel to New York and Miami just to shop (see O'Dougherty [1995]); in 1995 alone, 3 million Brazilians traveled abroad to shop, most of them to the United States (Veja 1425: 1, 44–50, January 3, 1996). Ninety-one percent of the Brazilians who visited the United States in 1995 declared that shopping was the primary purpose of their trip (Veja 1425: 47). Furthermore, Brazilian tourists spent the second-largest amount of money (after the Japanese) in 1995 in the United States and in France (Veja 1425: 49). Cosmopolitanism, particularly the demonstration of intimacy with European culture, is a symbolic strategy for social differentiation adopted by the elites and upwardly mobile strata. Identification with European culture is used in a number of ways, which sometimes include the allocation of Brazil to the "Western" (developed) sphere rather than to the developing sphere. This is illustrated by the statement of a medical doctor from Rio who, when asked about his views on Brazil's situation within the world system, referred to his recent experience in Florence, Italy (where the Interna-

tional Conference on AIDS had taken place), as having given him a chance to experience personally the Renaissance Italian art that had been part of his enculturation. As a Brazilian, he felt closer to Italy—and Europe, for that matter—than to the stereotypical developing countries of Africa and Asia.

2. See Minayo (1995).

3. Leblon is one of the middle- and upper-class oceanfront neighborhoods in Rio's Zona Sul, a "fancier" section of the city. Ipanema, Lagoa, Jardim Botânico, and Gávea are the other expensive neighborhoods that share the views of Lagoa Rodrigues de Freitas, Corcovado, and the Atlantic Ocean. Also within the Zona Sul are the once fancier beachfronts of Copacabana and Leme, as well as Botafogo, Flamengo, and, up the hill, Laranjeiras and Cosme Velho. Older neighborhoods occupy the downtown transition area (Glória, Santa Teresa, Lapa, Fátima, Centro) up to the Zona Norte, where there is a mixture of more conservative middle-class neighborhoods (such as Tijuca and Grajaú) and popular working-class and semi-suburban neighborhoods (such as São Cristóvão, Meier, Madureira, and Bom Sucesso), as well as a number of *favelas* (slums) such as Borel, Mangueira, and Andaraí. Favelas exist in most of the rocky hills scattered throughout the city, including Zona Sul, Pavão/Pavãozinho, Chapéu-Mangueira, Rocinha, Santa Marta, and Tabajaras. They create a sharp contrast to the middle-class *asfalto* (paved) neighborhoods.

4. Baixada Fluminense is an extensive *subúrbio* of Rio's North Zone. In Rio's urbanism, the periphery and suburban life is devaluated and left to those with lower incomes. The adjective *suburbano* (literally "suburban") is considered derogatory in carioca jargon.

5. Morel (1979).

6. Under the first military regimes, research activities were seen as subversive, and a few of the biomedical researchers from FIOCRUZ were deprived of their professional licenses. However, under the military president Médici, the government changed its policies toward FIOCRUZ and re-invested in it, allowing scientists to return and resume laboratory research activities.

7. This merit-oriented system hides the fact that in order to be admitted to public universities, candidate students go through severe entry exams, for which the average public education provides insufficient training. To improve their chances of being admitted, therefore, students must go to private schools and pay a higher tuition at the elementary and secondary level.

8. See, for example, Mann (1991).

9. See, for example, Harrington (1992).

10. See Daniel (1985), Daniel and Parker (1990, 1993), Guimarães et al. (1988), Ramos (1990).

11. Brasília has been the capital of Brazil since 1960. It was created from nothing on a central plateau in the "cerrado," a semi-desertic, arid land. Its creation reflects the megalomaniacal dreams of a new civilization that characterized Brazilian politicians after World War II. Rio, the former capital, kept its aristocratic glamour while losing to Brasília its decision-making and administrative power, along with federal funding.

12. Pedro Chequer, personal communication, Brasília, August 1989. I later discussed this issue with epidemiologist Maurício Peres and the AIDS program coordinator Celso Ramos Filho at the Federal University Hospital in Rio de Janeiro, in 1991. At that time, they had both been involved with the revisions of the standard case definition. The main piece summarizing the work had been submitted and

quoted a number of times, but it had not yet been accepted for publication. In 1993, a broader analysis that compared several AIDS case definitions with clinical cases was presented at the International Conference in Berlin by a group of doctors from the same unit. The *Lancet* published a synopsis of the work in the form of a letter to the editor (Durovni et al. [1993]).

13. An informal conversation with Brazilian immunologist Ana Faria in New York City helped me understand the scientific arguments against the limited character of contemporary models in virology and genetics and the use of a germ-theory approach to AIDS. Additional interviews with other Brazilian immunologists made me realize that a number of them were using an alternative paradigm, one that was different from the dominant paradigm of invasions, assaults, and counter-assaults, and quite different from the current obsession with molecular genetics (see Chapter 6). This does not mean that there is some type of "native" model for AIDS, but it suggests that some local scientists consider another paradigm to be at least as valid as the popular ones of the day. That paradigm appears to be based on the theories of Metchnikoff, who maintained that antibodies are not just "against" non-self but are produced within the self as well. I came across the model again in Alfred Tauber's (1994) work on the philosophy of immunology.

14. Hospital Evandro Chagas (1990: 1).

15. This part of the research was accomplished on the generous invitation of doctor Rosa Soares, who is also a professor of clinical medicine, whom I met at a training course on tropical medicine in FIOCRUZ. For a brief period, I was a member of her large team, which included residents and medical students under training, as well as the occasional specialist.

16. At a later time, I worked briefly with the epidemiology unit at the hospital Evandro Chagas. At that time, the team was trying to work with community organizations to evaluate a prevention program. The program, which was funded by WHO, was required for ethical reasons by a larger multi-center project. It was designed as a cohort study but was also intended for epidemiological surveillance, with the intention of following indicators that could be used in later studies of the efficacy of vaccines. This was not a part of my own research.

17. INAMPS, the centralized public health agency that owns and oversees several large hospitals, selected HU for the special AIDS Unit, which it assigned to DIP. At that time, INAMPS was coordinated by progressive officers (sometimes referred to as the "medical left") who were committed to public health and who endorsed the need to publicly respond to AIDS. DIP was committed to providing assistance and maintained the twelve (later eighteen) beds guaranteed for AIDS. The program hired eight specialists, most of them just out of the residency in DIP. The funds allocated to AIDS allowed them to obtain computers and set up laboratory facilities that would not have been possible otherwise.

18. The expression most commonly used to describe the health care system at the time of my fieldwork was *total sucateamento*, or "total trashing." That was during the presidency of Fernando Collor, whose vaguely neo-liberal projects included the shrinking of government-based services, even though there was no free market–based economic vitality to act as a counterpart.

19. Literally "tide slum," because of its proximity to water and its former layout of palafite constructions. Palafites have now been replaced by sturdier housing.

20. Caldas (1990: 17).

21. See Fraga (1990: 34).
22. Fraga (1990: 26).
23. Fraga (1990: 63).
24. Fraga (1990: 28–29).
25. Two cities overlap—the *favelas,* squatted up in the *morros* or huge stone hills that are spread all over Rio, and which are not legally available for urbanization; and the *asfalto,* or paved city, which is legally available for urbanization and is marked by the presence of urban housing, from high-rise luxury buildings to middle-income compounds, houses, or working-class *vilas.*
26. On top of the basic infrastructure of the University Hospital and its staff, the unit had a generous endowment provided by PETROBRAS, the national company for oil exploration and administration. That funding allowed the unit to create special laboratories to enhance the quality of diagnosis.
27. Some were actually hired and paid by DIP with funds from the special endowment to treat AIDS. They ceded to the services of their specialty, as DIP did not need them full time.
28. See Souza (1988).
29. I chose not to interview hospitalized patients, who were overburdened with the media's morbid eagerness to portray the *"aidético."* However, sometimes they wanted me to tape-record their statements. My interviews and informal interactions with people with AIDS in Brazil, though, happened mostly outside hospital settings.
30. Sometimes the "residual" and "all-purpose" character of social workers' knowledge pushed them to intervene when nobody else knew what to do. One social worker reported being called in to handle the case of a cross-gender patient who had the civil and biological identity of a male but the social and psychological identity of a female. No one could decide whether to assign the patient to the women's or the men's infirmary; the decision was left up to the social workers.
31. In this section of the book, the names of hospital personnel are pseudonyms. Recently, there has been a tendency to use the actual names of "informants" to acknowledge their contribution and often co-authorship. I adopted that procedure in the case of specific contributions by interviewed scientists (e.g., notes 36, 41, 44, 46, and 51 in Chapter 6).
32. Some chilling reports describe patients who had been abandoned by families that did not want them back. Most shocking for the social workers was that some mothers rejected their sons, contradicting the general belief in the solid nature of the mother/child bond. These cases were mentioned as exceptions rather than as the rule. More often, mothers, partners, and other family members were constantly around the unit. There was also the famous case of a patient who had been released from the hospital but could not be discharged because he had absolutely nowhere to go. His mother lived in the backlands of a northeastern state, far away from any type of medical care. He stayed in a hospital bed for months, even though he had no medical need for that. He even appealed for help from Princess Diana of Wales, who had once paid a visit to the hospital unit.
33. See, for example, Baker (1991), Kaplan (1991).
34. A similar situation is described by Peggy McGarrahan in a study of New York City nurses who treat people with AIDS (McGarrahan [1992]).
35. Some members of the hospital staff used stereotypical nuclear family metaphors to describe nurses as "mothers" (some were male), who handled the hardships of

domestic life, and doctors as "fathers" (the medical staff had a balanced female/male ratio), who were the "decision makers" but did not have to be there to handle the effect of those decisions. For a discussion of the medical decisions and the confidence needed to make them, see Bauman et al. (1991).

36. In Brazil, nursing is considered a professional/technical training rather than an academic one.

37. During my research, I could observe the consequences of and resistance to the introduction of new AIDS therapies, which involved extra work for the nurses. At some point during 1989 and 1990, the government distributed the equipment needed to administer aerosol pentamidine, a powerful prophylaxis for AIDS-related pneumonia (see Chapter 2). Its administration, however, involved a number of extra hours of nursing care and greater flexibility in house rules. A nurse with a special commitment to AIDS therapy might put in the extra energy and work for a while. His efforts were not mandatory, however. When he left, if no one else had an interest in it, or if no one else was available to administer this lifesaving procedure, it wouldn't get done. A doctor commented that had there been an extra incentive, such as greater involvement of the nursing staff in the overall program, things might not have become so difficult and contrary to the interests of the patient.

38. See, for example, Côrtes (1991).

39. The debate also took place in the United States. When an AIDS ward was planned for San Francisco General Hospital, the argument was that a new specialty was required to treat the simultaneous infections that arose in patients with AIDS. The counter-argument was that a separate ward would be like a leper-house and thus would reinforce the stigma that had been placed on the disease by the outside world (see Shilts [1987]).

40. Although that representation is not totally accurate—for instance, patients with hepatitis B return to the infectious disease ambulatory clinic many times and may neither be cured nor die—it is part of the inner core of beliefs or DIP myths that lead young medical students or residents to choose the specialty.

41. Female physicians treating people with AIDS tended to express the fear that "it could be me" more easily than their male colleagues, who, like most of the society, felt threatened by the sexual implications of being male and having AIDS. They were more likely to express the fear that "it could be my brother." As the number of women with AIDS in the wards increased and the stereotype of limited risk factors faded, health care professionals started to feel increasingly vulnerable—not because of the fear of contamination (which had occurred years before, with the very first cases, and which has been discussed in the international literature; see Dworkin et al. [1991]), but because they felt that they were close to the "epicenter" of the disease, being of the same age group as their patients and being sexually active. Deep in the subconscious and coming to consciousness every once in a while—whether in dreams, conversations, or therapy—that feeling and related fears were kept under control for better professional performance.

42. It should be noted that just before I started conducting my interviews (1991), there had been an incident that embittered many members of the team of health care professionals against one particular NGO and which extended to the whole category. A more self-empowered, college-educated member of GAPA–RJ was admitted to the emergency room and waited there for several days before he was given a regular hospital bed. Shocked by the poor conditions in the facility, his organization pro-

tested against the hospital in the newspapers. The health team was already tired of fighting for more resources to improve treatment. Upset about being targeted by an NGO, which should, they felt, be targeting the government for better health care, the doctors became distrustful of civil organizations, which they viewed as concerned with their own members' well-being much more than with general social causes.

43. Some used the expressive image of *morrer na praia* (literally, to die at the beach), meaning that they were almost there, reaching the end, only to miss it at the very last minute. It is also used to say that after so much work, and almost reaching your objectives, you would not make it through.

44. See Bolton and Orozco (1994).

45. The metaphor "the little brick in the wall" was often used by local researchers to describe their own participation in the production of the knowledge of AIDS. In contrast, a non-clinical researcher once described the goal of a scientist as being to "open windows in the walls," that is, to shed new light on the issues. She continued, "There are always people engaged in putting those little bricks that close the window entirely, so that you have to look for where to open a new window."

46. See, for example, Moraes and Carrara (1985a), Patton (1990). For a lengthier discussion, see Camargo (1994).

47. That included featuring anthropologist Richard Parker, a Berkeley graduate and professor in UERJ, on TV shows and in articles in popular magazines, as well as the rapid translation of his book *Bodies, Pleasures and Passions* (Parker [1991a]) into *Corpos, Prazeres e Paixões* (Parker [1991b]) by a popular press.

48. See, for example, *Globo,* November 5, 1987, and April 13, 1989, *Folha de São Paulo,* August 8, 1991, and October 3, 1991.

49. See Kalichman (1994).

6. War Metaphors in Germ Theory and Immunology

1. An illustration is the resident physician who punctured his finger with the blood of a terminally ill patient in Rio in 1991. He took AZT immediately, even though nobody knew whether it had any effect at that stage, or even if he was infected at all.

2. When I was in Rio, AZT was extremely expensive, even for the middle class (see, for example, Daniel 1991a, 1991b). Sometimes family and friends would have to pool their funds to purchase a bottle of Retrovir for a patient. When somebody died, his or her medicines were distributed among the needy through informal or NGO-based networks. Even the highly criticized office of Eduardo Côrtes in the AIDS Division (1990–1991) gave priority to the purchase of AZT for free distribution, going against a government that did not pay much attention to AIDS or want to spend money on patients whose cause it considered lost.

3. AZT was actually the initial basis for treatment activism in Brazil. Júlio Gaspar (1992) reported discussions in GAPA–SP as the motive for developing broader reflection, which materialized with *Cadernos Pela VIDDA* (see Chapter 4).

4. See Bader and Dorozynsky (1993), Jaret (1986), Latour (1984).

5. See Kruif (1926).

6. The German bacteriologist Koch is historically and medically as important as Pasteur; in fact, they were rivals at some point, exacerbating German and French nationalism.

7. Nussbaum (1990: 332).

8. The long-term results of the new combination drugs, which include the protease inhibitors, are still to be evaluated. Their short-term results evoked enthusiasm among doctors and the international AIDS community, but not without a few notes of skepticism: The "undetectable virus" may be hiding somewhere in the body, and the use of these drugs may be a step toward the creation of new variants of multi-resistant virus.

9. Sontag (1979).

10. Crimp (1987), Haraway (1989), Martin (1990, 1994), Patton (1985, 1990), Sontag (1989), Threichler (1987, 1988, 1989, 1992).

11. Kuhn (1962).

12. Martin (1994).

13. E.g., Jaret (1986).

14. Martin (1994).

15. Coutinho et al. (1984), Varela (1979, 1988), Varela and Coutinho (1991), Varela et al. (1988), Vaz and Varela (1978).

16. Martin (1994: 110).

17. 1992, personal communication.

18. Tauber (1994), Vaz and Faria (1992); see also Kamminga (1994).

19. See Tauber (1994), Vaz and Faria (1992).

20. See, for example, Thomas (1974).

21. This has been documented for the Fore of New Guinea by Shirley Lindenbaum (1979).

22. See Tauber (1994).

23. At the 1994 AAA conference in Atlanta, anthropologist Emily Martin provided a number of examples of the current trends for immune system diseases such as multiple sclerosis and arthritis, which included the ingestion of bone components that are generally not part of the human diet. That type of data confirmed what the immunologist Nelson Vaz claimed to be the reason for his epistemological split from classical immunology: the long-documented and observed phenomenon of oral immunization achieved by certain animals that eat certain plants. Vaz (1992, personal communication) overcame that epistemological crisis by adopting an epistemology of dynamic systems, as expressed by the philosopher of immunology Francisco Varela (1979).

24. Moreira (1974: 39).

25. See Delaporte (1991), Moreira (1974).

26. Quoted by Moreira (1974: 73).

27. Falcão (1973).

28. Besides yellow fever, Cruz led a campaign against bubonic plague that included the extermination of city rats, and a campaign against smallpox, using vaccine serum produced in the institute. The latter generated the insurrections known as *revolta da vacina* (vaccine revolt), a mass response against the harsh policies of public health officers (see Carvalho [1987]).

29. See Benchimol (1990).

30. Manguinhos was also a must for celebrity visitors. House documents include photographs of visits by Albert Einstein and other major scientists, as well as royalty and politicians from around the world.

31. See Thielen et al. (1991).

32. See Chagas (1990: 84).

33. Chagas (1990: 85). *P. carinii* was still described as a protozoan within the context of the AIDS epidemic, and *P. carinii* pneumonia was often referred to as a rare disease that affected mainly infants and the elderly. Only some years after the AIDS epidemic began was *P. carinii* reclassified as a fungus. For a detailed description, see Bastos (1996).

34. Benchimol (1990), Moreira (1974: 157).

35. See Stepan (1976: 120).

36. The latter characterization of Chagas disease is far more complex. It is described as a slowly progressive disease that involves cardiac pathologies. An interesting recent finding is that the damage induced by Chagas disease is caused by the lymphocytes of the infected person. The patient has an autoimmune response to the "disguised" lymphocytes and is not responding directly to the action of an infectious agent (Ribeiro dos Santos [1992, personal communication]).

37. For a detailed account of the multiple aspects of Chagas disease, see Briceño-Léon (1990, 1993).

38. See Chagas (1990: 187–230).

39. Chagas (1990: 187).

40. Benchimol (1990).

41. Garcia (1992, personal communication).

42. Stepan (1976: 122).

43. See, for example, Benchimol (1990), Benchimol and Teixeira (1993), Britto (1995), Thielen et al. (1991).

44. Insights into the social determinants of malaria and other "tropical diseases" were provided by Dr. Maurício Peres from the Nucleus of Studies in Collective Health, UFRJ; Dr. Paulo Chagastelles Sabroza from the National School of Public Health, FIOCRUZ; and Dr. Celso Ramos Filho from the University Hospital, UFRJ. I also attended a number of conferences while conducting this research project. Errors or inadequate representations thereof in my presentation of the issue are my own responsibility.

45. Actually, cholera may be considered an example of the rapid response of socially oriented biomedical research institutions such as FIOCRUZ. The South American cholera epidemic started during my fieldwork in Rio, and was the topic of much speculation by the media and even the government. It was estimated that the number of cholera cases in Brazil would reach the hundreds of thousands if no action were taken. The minister of health, Alceni Guerra, used the opportunity to create a quite original campaign involving barefoot educators in the countryside, but this was no more than a pretext for the purchase of large amounts of unnecessary goods, such as bicycles and umbrellas, for his financial gain. He was dismissed as a consequence of these purchases, which were made at a store in the city of Curitiba, Paraná, which was owned by several of his political friends. Independently, a team from FIOCRUZ that had been involved in cholera research went to the Peru–Brazil border to evaluate the situation and make plans for immediate action. They understood that the water route from feces to ingestion that had been responsible for the propagation of Vibrion cholera was different on the Brazilian and the Andean sides of the border. In Brazil, more widespread sanitation helped prevent the catastrophic spread of the epidemic. So instead of the predicted 200,000 cases, only 20,000 were recorded.

46. Fujimura (1988), Vaz (1992, personal communication).

47. In these countries, the less attractive field of infectious diseases was limited mostly to military medicine or former colonial institutes. Before the onset of AIDS, the field of immunology was largely associated with transplants or treating people with low-key allergies. Promising careers were pursued in other fields, such as the molecular biology of cancer or gene mapping.

48. See Garrett (1992).

49. See Haraway (1988).

50. The polymerase chain reaction (PCR) is a test based on duplicating DNA segments that is used to determine the amino acid sequence of a protein or nucleic acid. It is used to determine whether a specific virus is present. The PCR became popular in HIV research, because unlike antibody-based ELISA and the Western Blot tests, a PCR can verify whether a specific virus (not the immune responses to it) is present.

51. This was a real case, witnessed in the city of Salvador in Bahia, in 1991. Dr. Bernardo Galvão-Castro, who moved from FIOCRUZ in Rio to organize a special and well-equipped research laboratory in Bahia, shared with me the contradictions and distress experienced by researchers who work in such fringe environments.

7. *Conclusions*

1. See, for example, Latour (1987), Pinch (1992).

2. Duesberg (1987, 1989, 1992, 1994) and followers dismissed the role of HIV; Root-Bernstein (1993) blamed the rapid and sloppy process of achieving consensus on AIDS; Coulter (1987) saw it as another form of syphilis; the editors of the *Spheric* (1994) called AIDS a "hoax"; and the *New York Native* consistently dismissed the term by putting it in quotes ("AIDS").

3. See, for example, Latour and Wolgar (1979), Latour (1987), Knorr-Cetina (1981).

4. See Kuhn (1962) and Mulkay (1972).

5. See, for example, Kritsky (n.d.).

6. Dr. Paulo Teixeira (1992, personal communication).

References

American Anthropological Association. 1993. Correspondence. *Anthropology Newsletter* 34(5).

Associação Brasileira Interdisciplinar de AIDS. 1988a. *ABIA: The Brazilian Response to the AIDS Challenge Based on Solidarity*. Rio de Janeiro: ABIA.

Associação Brasileira Interdisciplinar de AIDS. 1988b. AIDS no Brasil: Incidência e evidência. In AIDS: Somos todos mortais. *Comunicações do ISER* 31: 4–8.

Associação Brasileira Interdisciplinar de AIDS. 1988c. Relatório inicial: Impacto social da AIDS no Brasil. *Boletim Abia* 1: 2–3.

Associação Brasileira Interdisciplinar de AIDS. 1988d. Quem semeia pânico, colhe epidemia: Caras e máscaras de uma campanha equivocada. *Boletim Abia* 2: 1.

Associação Brasileira Interdisciplinar de AIDS. 1988e. Onze críticas a uma campanha desgovernada. *Boletim Abia* 2: 2.

Associação Brasileira Interdisciplinar de AIDS. 1988f. AIDS: O número de casos e o caso dos números. *Boletim Abia* 3: 1.

Associação Brasileira Interdisciplinar de AIDS. 1988g. Números em pauta. *Boletim Abia* 3: 2.

Associação Brasileira Interdisciplinar de AIDS. 1988h. Sangue novo. *Boletim Abia* 4: 1.

Associação Brasileira Interdisciplinar de AIDS. 1988i. Uma ou duas coisas que sabemos sobre o sangue. *Boletim Abia* 4: 4–5.

Associação Brasileira Interdisciplinar de AIDS. 1988j. E onde fica o Pinto Fernandes? *Boletim Abia* 5: 1–2.

Associação Brasileira Interdisciplinar de AIDS. 1989a. AIDS: Desafios e armadilhas. *Boletim Abia* 6: 1.

Associação Brasileira Interdisciplinar de AIDS. 1989b Encontro de ONGs em Montreal. *Boletim Abia* 6: 12.

Associação Brasileira Interdisciplinar de AIDS. 1989c. E se o Presidente? *Boletim Abia* 8: 1–2.

Associação Brasileira Interdisciplinar de AIDS. 1989d. A vida em emergência: Rede Brasileira de Solidariedade, Porto Alegre, 1989. *Boletim Abia* 9: 1–2.

Associação Brasileira Interdisciplinar de AIDS. 1990a. O boicote à VI Conferência Internacional sobre AIDS: Um apelo a todas as pessoas e organizações que trabalham com AIDS. *Boletim Abia* 10: 12.

Associação Brasileira Interdisciplinar de AIDS. 1990b. A VI Conferência Internacional sobre AIDS e os direitos humanos. *Ação Anti-AIDS* 9: 7–10.

Associação Brasileira Interdisciplinar de AIDS. 1991a. Prevenção fatal. *Boletim Abia* 13: 1–2.

Associação Brasileira Interdisciplinar de AIDS. 1991b. Proteste você também. *Boletim Abia* 13: 12.

Associação Brasileira Interdisciplinar de AIDS. 1991c. AIDS: Prioridade nacional de saúde. *Boletim Abia* 14: 1–2.

Associação Brasileira Interdisciplinar de AIDS. 1991d. A solidariedade é uma grande empresa. *Boletim Abia* 14: 10–11.

References

Associação Brasileira Interdisciplinar de AIDS. 1992. *AIDS no Mundo em 1992.* Boletim especial (special issue).

Associação Brasileira Interdisciplinar de AIDS. 1993a. *Por uma nova estratégia de saúde frente à AIDS: Uma proposta da Coalizão Global de Políticas Contra a AIDS para os desafios dos anos 90.* Boletim especial (special issue).

Associação Brasileira Interdisciplinar de AIDS. 1993b. Projeto do Banco Mundial está ameaçado. *Boletim Abia* 21: 12.

Associação Brasileira Interdisciplinar de AIDS. 1994. *Projeto Homossexualidades: Ano e meio de trabalho.* Boletim especial (special issue).

Abramson, Paul R. 1992. Sex, Lies, and Ethnography. In *The Time of AIDS*, ed. Gilbert Herdt and Shirley Lindenbaum, 101–123. London: Sage.

ACT UP/NY (AIDS Coalition to Unleash Power, New York) Women & AIDS Book Group. 1990. *Women, Aids and Activism.* Boston: South End Press.

Aggleton, Peter, Graham Hart, and Peter Davies, eds. 1989. *AIDS: Social Representations, Social Practices.* London: Falmer Press.

Aggleton, Peter, Peter Davies, and Graham Hart, eds. 1990. *AIDS: Individual, Cultural, and Policy Dimensions.* London: Falmer Press.

Aggleton, Peter, Graham Hart, and Peter Davies, eds. 1991. *AIDS: Responses, Interventions, and Care.* London: Falmer Press.

Aggleton, Peter, Peter Davies, and Graham Hart, eds. 1992. *AIDS: Rights, Risk, and Reason.* London: Falmer Press.

Aggleton, Peter, Peter Davies, and Graham Hart, eds. 1993. *AIDS: Facing the Second Decade.* London: Falmer Press.

AHRTAG (Appropriate Health Resources and Technologies Action Group). 1990a. Por que as organizações não participarão da maior conferência internacional de AIDS. *Ação Anti-AIDS* 9: 1, 12.

AHRTAG (Appropriate Health Resources and Technologies Action Group). 1990b. Políticas de Solidariedade. *Ação Anti-AIDS* 12: 1, 11.

Albrow, Martin, and Elizabeth King, eds. 1990. *Globalization, Knowledge and Society: Readings from International Sociology.* London: Sage.

Almeida Filho, Naomar. 1989a. *Epidemiologia Sem Números: Uma Introdução Crítica à Epidemiologia.* Rio de Janeiro: Campus.

Almeida Filho, Naomar. 1989b. The Object of Knowledge in Epidemiology. Paper presented at the Seminar on Theoretical Aspects of Epidemiological Science, University of North Carolina School of Public Health, Chapel Hill.

Almeida Filho, Naomar. 1990a. Paradigms in Epidemiology. Paper presented at the roundtable Theoretical Challenges for Epidemiology, First Brazilian Conference on Epidemiology, Campinas.

Almeida Filho, Naomar. 1990b. Integration of Qualitative and Quantitative Methodology in Epidemiologic Research. Lecture at the First International Conference in Ethnicity and Disease, Case Western Reserve University, Cleveland.

Almeida Filho, Naomar. 1991a. A desconstrução do conceito de risco. Lecture at the seminar Conceptual Advances in Epidemiology: Challenges of the 1990s. Rio de Janeiro: National School of Public Health.

Altman, Dennis. 1987. *AIDS in the Mind of America: The Social, Political, and Psychological Impact of a New Epidemic.* Garden City, N.Y.: Anchor Press.

Altman, Dennis. 1988. Legitimization through Disaster: AIDS and the Gay Movement. In *AIDS: The Burdens of History,* ed. Elizabeth Fee and Daniel Fox, 301–315. Berkeley: University of California Press.

Altman, Dennis. 1995. *Poder e Comunidade: Respostas Organizacionais e Culturais à AIDS*. Rio de Janeiro: ABIA/IMS/Relume–Dumará.
Amat-Roze, Jeanne-Marie. 1993. Les inegalités geographiques de l'infection à VIH et du SIDA en Afrique Sud-Saharienne. *Soc Sci Med* 36(10): 1247–1256.
American Academy of Arts and Sciences. 1986. America's Doctors, Medical Science, Medical Care. *Daedalus* 115(2) (special issue).
Amin, Samir. 1976. *Unequal Development: An Essay on the Social Formations of Peripheral Capitalism*. New York: Monthly Review Press.
Amin, Samir. 1989. *Eurocentrism*. New York: Monthly Review Press.
Ankrah, Maxine, et al. 1994. Women and Children and AIDS. In *AIDS in Africa*, ed. Max Essex et al., 533–546. New York: Raven Press.
Arenas, Reynaldo. 1993. *Before the Night Falls*. New York: Viking.
Arno, Peter. 1991 [1989]. An Expanded Role for Community-Based Organizations. In *The AIDS Reader*, ed. Nancy MacKenzie, 497–504. New York: Meridian.
Arno, Peter S., and Karyn L. Feiden. 1992. *Against the Odds: The Story of AIDS Drug Development, Politics and Profits*. New York: HarperCollins.
Arnold, David, ed. 1988. *Imperial Medicine and Indigenous Societies*. Manchester: Manchester University Press.
Arnold, David, ed. 1993. *Colonizing the Body: State Medicine and Epidemic Disease in Nineteenth-Century India*. Berkeley: University of California Press.
As Sy, El Hajd. 1995. Networking for Mutual Support. *Exchanges* 1995(1): 1–2.
Bader, Jean-Michel, and Alexandre Dorozynski. 1993. La revanche des microbes, première partie: La nouvelle menace bactérienne. *Science & Vie* 904: 49–62.
Baker, G. H. B. 1991. Psychological factors and immunity: a selective review of recent advances. *Psychiatria Fennica* 22: 47–52.
Baldo, M., and A. J. Cabral. 1991. Low-intensity wars and social determination of HIV transmission: the search for a new paradigm to guide research and control of the HIV/AIDS pandemic. In *Action of AIDS in Southern Africa: Maputo Conference on Health in Transition in Southern Africa, April 1990*, ed. Z. Stein and A. Zwi. New York: Committee for Health in Southern Africa.
Baracca, Angelo, and Arcangelo Rossi. 1976. *Marxismo e scienze naturali: Per una storia integrale delle scienze*. Bari, Italy: De Donato.
Barber, Bernard. 1990. *Social Studies of Science*. London: Transaction.
Barletta, Giuseppe. 1978. *Marxismo e teoria della scienza: Materiali di analisi*. Bari, Italy: Dedalo libri.
Barnes, Barry. 1974. *Scientific Knowledge and Sociological Theory*. London: Routledge and Kegan Paul.
Barnes, Barry. 1982. *T. S. Kuhn and Social Science*. New York: Columbia University Press.
Barnes, Barry. 1988. *The Nature of Power*. Cambridge, U.K.: Polity Press.
Barnes, Barry, and Steven Shapin. 1979. *Natural Order: Historical Studies of Scientific Culture*. London: Sage.
Barnes, Barry, and David Edge, eds. 1982. *Science in Context: Readings in the Sociology of Science*. Milton Keynes, U.K.: The Open University Press.
Barnett, Richard, and John Cavanagh. 1994. *Global Dreams: Imperial Corporations and the New World Order*. New York: Simon and Schuster.
Bastide, Roger, and Florestan Fernandes. 1971. *Brancos e Negros em São Paulo*. São Paulo: Companhia Editora Nacional.
Bastos, Cristiana. 1996. The Metamorphosis of the Germ: *Pneumocystis carinii* in Two (Medical) Cultures. Paper presented at the session "On Animation and Cessa-

tion," 95th Annual Meeting of the American Anthropological Association, San Francisco.

Bateson, Mary Catherine, and Richard Goldsby. 1988. *Thinking AIDS: The Social Response to the Biological Threat*. New York: Addison–Wesley.

Bauman, Andrea O., Raise B. Deber, and Gail G. Thompson. 1991. Overconfidence among physicians and nurses: the micro-certainty, macro-uncertainty. *Soc Sci Med* 32(2): 167–174.

Bayer, Ronald. 1985. AIDS and the gay community: between the specter and the promise of medicine. *Social Research* 52(3): 581–606.

Beloqui, Jorge. 1994. Produto vacinal será testado no Brasil. *Boletim Abia* 23: 9.

Benchimol, Jaime L. 1995. Domingos José Freire e os Primórdios da Bacteriologia no Brasil. *História, Ciências, Saúde, Manguinhos* 2(1): 67–98.

Benchimol, Jaime L., ed. 1990. *Manguinhos, do sonho à vida: A ciência na Belle Epoque*. Rio de Janeiro: Casa de Oswaldo Cruz, FIOCRUZ.

Benchimol, Jaime Larry, and Luis Antônio Teixeira. 1993. *Cobras, lagartos e outros bichos: Uma história comparada dos Institutos Oswaldo Cruz e Butantan*. Rio de Janeiro: Editora UFRJ.

Bergensen, Albert. 1990. Turning world-system theory on its head. *Social Studies of Science* 7(23): 67–81.

Berger, Peter, and Thomas Luckman. 1966. *The Social Construction of Reality*. New York: Doubleday.

Bernal, J. D. 1972. *The Social Function of Science*. Cambridge: MIT Press.

Berridge, Virginia, and Philip Strong. 1991. AIDS and the relevance of history (review article). *Social History of Medicine* 4(1): 129–138.

Berridge, Virginia, and Philip Strong, eds. 1993. *AIDS and Contemporary History*. Cambridge: Cambridge University Press.

Bloor, David. 1983. *Wittgenstein: A Social Theory of Knowledge*. New York: Columbia University Press.

Bloor, David. 1986. 1971 Essay review: two paradigms for scientific knowledge. *Science Studies* 101–115.

Bloor, David. 1991. *Knowledge and Social Imagery*. London: Routledge and Kegan Paul.

Bolton, Ralph. 1992. Mapping Terra Incognita: Sex Research for AIDS Prevention—An Urgent Agenda for the 1990s. In *The Time of AIDS*, ed. Gilbert Herdt and Shirley Lindenbaum, 124–158. London: Sage.

Bolton, Ralph. 1995. Coming Home: The Journey of a Gay Ethnographer in the Years of the Plague. In *Gay and Lesbian Fieldwork and Ethnography*, ed. E. Lewin and William Leap. Champaign: University of Illinois Press (in press).

Bolton, Ralph, ed. 1989. *The AIDS Pandemic: A Global Emergency*. New York: Gordon and Breach.

Bolton, Ralph, and Merryl Singer, eds. 1992. *Rethinking AIDS Prevention: Cultural Approaches*. New York: Gordon and Breach.

Bolton, Ralph, and Gail Orozco. 1994. *The AIDS Bibliography*. Arlington, Va.: American Anthropological Association.

Boston Women's Health Book Collective. 1973. *Our Bodies, Ourselves*. New York: Simon and Schuster.

Botelho, António José Junqueira. 1990. The Professionalization of Brazilian scientists: the Brazilian society for the progress of science (SBPC) and the state, 1948–60. *Social Studies of Science* 20: 473–502.

Bouchara, Jacques. 1991. Percalços da solidariedade. [Hazards of Solidarity]. *Boletim Abia* 14: 9–10.

Bourdieu, Pierre. 1975. The specificity of the scientific field and the social conditions of the progress of reason. *Social Science Information* 14: 19–47.

Bowser, Benjamin P. 1994. HIV Prevention and African Americans: A Difference of Class. In *AIDS Prevention and Services: Community Based Research,* ed. Johannes Van Vugt, 93–108. Westport, Conn.: Bergin and Garvey.

Braiterman, Jared. 1991a. Brazilian government funds explicit AIDS education booklets. *Advocate* 576: 60–63.

Braiterman, Jared. 1991b. A sex-positive sex worker-positive AIDS project in Brazil. *Gay Community News,* June 23–29.

Brandt, Alan M. 1987. *No Magic Bullet: A Social History of Venereal Disease in the United States since 1880. With a new chapter on AIDS.* Oxford: Oxford University Press.

Brante, Thomas, Steve Fuller, and William Lynch, eds. 1993. *Controversial Science: From Context to Contentions.* Albany: State University of New York Press.

Breihl, Jaime. 1981. *Epidemiología: Economía, Medicina, y Política.* Santo Domingo: SESPAS.

Briceño-Leon. 1990. *La Casa Enferma: Sociología de la Enfermedad de Chagas.* Caracas: Fondo Editorial Acta Cientifica Venezolana.

Briceño-Leon. 1993. Social Aspects of Chagas Disease. In *Knowledge, Power and Practice: The Anthropology of Medicine and Everyday Life,* ed. Shirley Lindenbaum and Margaret Lock, 287–300. Berkeley: University of California Press.

Britto, Nara. 1995. *Oswaldo Cruz: A Construção de um Mito na Ciência Brasileira.* Rio de Janeiro: Fiocruz.

Brown, James Robert. 1989. *The Rational and the Social.* London: Routledge.

Burdick, John. 1995. The Eyes of Anastacia: Color and Hegemony in a Brazilian Religious Devotion. Paper presented at the 19th International Congress of the Latin American Studies Association, Washington, D.C.

Burkett, Elinor. 1995. *The Gravest Show on Earth: America in the Age of AIDS.* New York: Houghton Mifflin.

Burr, Chandler. 1995. In search of the gay gene. *Advocate* 697 (December 26): 1, 36–42.

Cagnon, John H. 1988. Sex research and sexual conduct in the age of AIDS. *Journal of Acquired Immune Deficiency Syndromes* 1(6): 593–601.

Cagnon, John H. 1990. Disease and Desire. In *Living with AIDS,* ed. Stephen Graubard, 181–211. Cambridge, Mass.: MIT Press.

Caldwell, John C. 1995. Understanding the AIDS epidemic and reacting sensibly to it. *Soc Sci Med* 41(3): 299–302.

Caldwell, John C., I. O. Orubuloye, and Pat Caldwell. 1992. Underreaction to AIDS in subsaharian Africa. *Soc Sci Med* 34(11): 1169–1182.

Caldas, Renato. 1990. Prefácio. In *A Implantação do Hospital Universitário da UFRJ (1974/1978),* by Clementino Fraga Filho. 17–19. Rio de Janeiro: Fundação Universitária José Bonifácio.

Callen, Michael, ed. 1987. *Surviving and Thriving with AIDS.* New York: People with AIDS Coalition.

Callen, Michael, ed. 1990. *Surviving AIDS.* New York: HarperCollins.

Camargo, Kenneth Rochel de, Jr. 1994. *As ciências da AIDS & a AIDS das ciências: O discurso médico na construção da AIDS.* Rio de Janeiro: ABIA/IMS/Relume–Dumará.

Caplan, Pat. 1987. *The Cultural Construction of Sexuality.* London: Routledge.

Carballo, M. 1988. International Agenda for AIDS Behavioral Research. In *AIDS 1988: AAAS Symposia Papers,* ed. R. Kulstad, 271–273. Washington, D.C. American Association for the Advancement of Science.

Carballo, M., John Cleland, Michel Caral, and Gary Abrecht. 1989. Research agenda: a cross-national study of patterns of sexual behavior. *Journal of Sex Research* 26: 287–299.

Cardoso, Fernando Henrique, and Enzo Faletto. 1967. *Dependencia y desarrollo en America Latina: Ensayo de interpretation sociologica.* Mexico: Siglo Veinteuno.

Carlson, Robert G., Harvey Siegal, and Russel S. Falck. 1994. Ethnography, Epidemiology, and Public Policy: Needle-Use Practices and HIV Risk Reduction among Injecting Drug Users in the Midwest. In *Global AIDS Policy,* ed. Douglas Feldman, 185–214. Westport, Conn.: Bergin and Garvey.

Carrier, Joseph. 1995. *De los Otros: Intimacy and Homosexuality among Mexican Men.* New York: Columbia University Press.

Carrier, Joseph, and Raúl Magaña. 1992. Use of Ethnosexual Data on Men of Mexican Origin for HIV/AIDS Prevention Programs. In *The Time of AIDS,* ed. Gilbert Herdt and Shirley Lindenbaum, 243–258. London: Sage.

Carter, Erica, and Simon Watney, eds. 1989. *Taking Liberties: AIDS and Cultural Politics.* London: Serpent's Tail.

Carvalho, José Murilo de. 1987. Os Bestializados: *O Rio de Janeiro e a República que não foi.* São Paulo: Companhia das Letras.

Cash, Kathleen, Bupa Anasuchatkul, and Wartana Busayawory. 1994. "Lamyai" teaches young Thai women about AIDS and STDs. *Exchanges* 1994(1): 4–7.

CDC (Centers for Disease Control). 1981a. Pneumocystis Pneumonia—Los Angeles. *MMWR* 30: 250–252.

CDC (Centers for Disease Control). 1981b. Kaposi's sarcoma and pneumocystic pneumonia among homosexual men—New York City and California. *MMWR* 30: 305–308.

Center for Social Research and Understanding. 1991. Radical experiments: social movements take on technoscience. *Socialist Review* 21(2).

Chagas Filho, Carlos. 1993. *Meu Pai.* Rio de Janeiro: Casa de Oswaldo Cruz, Fundação Oswaldo Cruz.

Chirimuuta, Richard C., and Rosalind J. Chirimuuta. 1989. *AIDS, Africa, and Racism.* London: Free Association Books.

Chubin, D. E., and K. E. Studer. 1978. The politics of cancer. *Theory and Society* 6: 55–74.

Chubin, Daryl E., and Sal Restivo. 1983. The Mooting of Science Studies: Research Programmes and Science Policy. In *Science Observed: Perspectives on the Social Study of Science,* ed. Karin Knorr-Cetina and Michael Mulkay, 52–85. London: Sage.

Clarck, Norman. 1985. *The Political Economy of Science and Technology.* Oxford: Basil Blackwell.

Clarke, Adele, and Joan Fujimura, eds. 1992. *The Right Tools for the Right Job: At Work in Twentieth Century Life Sciences.* Princeton, N.J.: Princeton University Press.

Clatts, Michael. 1991. Homeless youth and AIDS: challenge for an anthropological practice. *AIDS and Anthropology Bulletin* 3(2): 8–10.

Clatts, Michael. 1993. Poverty, Drug Use and AIDS: Converging Issues in the Life Stories of Women in Harlem. In *Wings of Gauze: Women of Color and Experience of*

Health and Illness, ed. B. Bair and S. Cayleff, 328–339. Detroit: Wayne State University Press.
Clatts, Michael C., W. R. Davies, Sherry Deren, Douglas S. Goldsmith, and Stephanie Tortu. 1994. AIDS Risk Behavior among Drug Injectors in New York City: Critical Gaps in Prevention Policy. In *Global AIDS Policy,* ed. Douglas Feldman, 215–235. Westport, Conn.: Bergin and Garvey.
Cohn, Amélia. 1995. NGOs, Social Movements, and the Privatization of Health Care: Experiences in São Paulo. In *New Paths to Democratic Development in Latin America: The Rise of NGO–Municipal Collaboration,* ed. Charles Reily, 85–98. Boulder, Colo.: Lynne Rienner.
Collins, H. M. 1982. *Sociology of Scientific Knowledge: A Sourcebook.* Bath, Avon: Bath University Press.
Collins, H. M. 1983. The sociology of scientific knowledge: studies of contemporary science. *Annual Review of Sociology* 9: 265–285.
Collins, H. M. 1985. *Changing Order: Replication and Induction in Scientific Practice.* London: Sage.
Connors, Margaret M. 1992. Risk perception, risk taking and risk management among intravenous drug users: implications for AIDS prevention. *Soc Sci Med* 34(6): 591–601.
Cooper, Charles. 1973. *Science, Technology and Development: The Political Economy of Technological Advance in Underdeveloped Countries.* London: Frank Lass.
Côrtes, Eduardo. 1991. A luta da Aids. *Jornal do Brasil,* May 3, 1991.
Costa, Dina Czeresnia, ed. 1990. *Epidemiologia: Teoria e objecto.* São Paulo: HUCITEC–ABRASCO.
Costa, Dina Czeresnia, ed. 1993. Construção científica e inovação teórica: Um desafio para a epidemiologia. *Physis, Revista de Saúde Coletiva* 3(1): 77–90.
Coulter, Harry. 1987. *AIDS and Syphilis: The Hidden Link.* Berkeley: North Atlantic Books.
Coutinho, António, L. Forni, D. Holmberg, F. Ivars, and N. Vaz. 1984. From an antigen-centered, clonal perspective of immune responses to an organism-centered, network perspective of autonomous activity in a self-referential immune-system. *Immunol Rev* 79: 151–168.
Crapanzano, Vincent. 1995. *The Moment of Prestidigitation: Magic, Illusion, and Mana in the Thought of Emile Durkheim and Marcel Mauss.* Stanford: Stanford University Press (in press).
Crawford, Robert. 1994. The boundaries of the self and the unhealthy other: reflections on health, culture and AIDS. *Soc Sci Med* 38(10): 1347–1365.
Crick, Malcolm R. 1982. Anthropology of knowledge. *Annual Review of Anthropology* 11: 287–313.
Crimp, Douglas, ed. 1987. AIDS: Cultural analysis. *Cultural Activism* 43 (special issue) (Winter 1987). Cambridge, Mass.: MIT Press.
Crimp, Douglas, with Adam Rolston. 1990. *AIDS-Demo-Graphics.* Seattle: Bay Press.
Crosby, A. 1976. *Epidemic and Peace, 1918.* Westport, Conn.: Greenwood Press.
Cueto, Marcus. 1989. Andean biology in Peru: scientific styles on the periphery. *Isis* 80: 640–658.
Daly, John A. 1990. Comments on Turnbull's push for a malaria vaccine. *Social Studies of Science* 20: 371–379.
Daly, John A. 1991. Does a constructivist view require epistemological relativism? A response to Turnbull. *Social Studies of Science* 21: 568–71.

Daniel, Herbert. 1985. A síndrome do preconceito. *Comunicações do ISER* 17: 48–56.

Daniel, Herbert. 1989. *Vida Antes da Morte/ Life before Death.* [Bilingual edition.] Rio de Janeiro: Jaboti.

Daniel, Herbert. 1990. O primeiro AZT a gente nunca esquece. *Boletim Abia* 11: 11.

Daniel, Herbert. 1991a. A Aids do governo. *Ultima Hora*, RJ, February 25, 1991.

Daniel, Herbert. 1991b. A doença da burocracia. *Jornal do Brasil*, June 6, 1991.

Daniel, Herbert. 1993. *Sexuality, Politics and AIDS in Brazil.* London: Falmer Press.

Daniel, Herbert, and Richard Parker. 1990. *AIDS, A Terceira Epidemia: Ensaios E Tentativas.* São Paulo: Iglu.

Danziger, Renee. 1994. The Social Impact of HIV/AIDS in Developing Countries. *Soc Sci Med* 39(7): 905–907.

Davies, Peter. 1989. Some Notes on the Structure of Homosexual Acts. In *AIDS: Social Representations, Social Practices,* ed. Peter Aggleton et al., 147–159. London: Falmer Press.

Davies, Peter, and T. Coxon. 1990. Patters in Homosexual Relations: The Use of the Diary Method. In *Sexual Behaviour and Risk of HIV Infection: Proceedings of an International Workshop Supported by the European Communities,* ed. Michel Hubert, 59–78. Brussels: Publications des Facultés Universitaires Saint-Louis.

Davies, Peter, et al. 1993. *Sex, Gay Men and AIDS.* London: Falmer Press.

Davis, D. L., and R. G. Whitten. 1987. The cross-cultural study of human sexuality. *Annual Review of Anthropology* 16: 69–.

Dedijer, Stevan. 1968. Underdeveloped Science in Underdeveloped Countries. In *Criteria for Scientific Development,* ed. Edward Shils, 13–163.

Delaporte, François. 1991. *The History of Yellow Fever. An Essay on the Birth of Tropical Medicine.* [Foreword by Georges Canguillen. Translated by Arthur Goldhammer.] Cambridge, Mass.: MIT Press.

Dhillon, H. S., and Lois Philip. 1994. *Health Promotion and Community Development Action for Health in Developing Countries.* Geneva: WHO.

Dubos, René. 1965. *Man Adapting.* New Haven, Conn.: Yale University Press.

Duesberg, Peter. 1987. Retroviruses as carcinogens and pathogens: expectations and reality. *Cancer Res* 47: 1199–1220.

Duesberg, Peter. 1989. Human immunodeficiency virus and acquired immunodeficiency syndrome: correlation but not causation. *Proc Natl Acad Sci USA* 86: 755–764.

Duesberg, Peter. 1992. AIDS acquired by drug consumption and other noncontagious risk factors. *Pharmacol Ther* 55: 201–277.

Duesberg, Peter. 1994. Infectious AIDS: stretching the germ theory beyond its limits. *Int Arch Allery Immunol* 10: 118–126.

Dumit, Joseph. 1995. Twenty-first Century PET: Looking for Mind and Morality through the Eye of Technology. In *Technoscientific Imaginaries,* ed. George Marcus, 87–128. Chicago: University of Chicago Press.

Durkheim, Emile. 1912. *Les Formes Elémentaires de la Vie Religieuse: Le Système Totémique en Australie.* Paris: Alcan.

Durkheim, Emile, and Marcel Mauss. 1963. *Primitive Classification.* Chicago: University of Chicago Press.

Durovni, B., M. Pinto, and M. Schechter. 1993. AIDS case definitions in developing countries (letter; with comment). *Lancet* 342(8878): 1054.

Dworkin, Joan, Gary Albrecht, and Judith Cooksey. 1991. Concern about AIDS among hospital physicians, nurses and social workers. *Soc Sci Med* 33(3): 239–248.

Earickson, Robert J. 1990. International behavioral responses to a health hazard: AIDS. *Soc Sci Med* 31(9): 951–962.

Edwards, Jeanette, S. Franklin, E. Hirsch, F. Price, and M. Strathern, eds. 1993. *Technologies of Procreation: Kinship in the Age of Assisted Conception*. Manchester: Manchester University Press.

Ehrenreich, Barbara. 1973. *Complaints and Disorders: The Sexual Politics of Sickness*. Old Westbury, N.Y.: Feminist Press.

Ehrenreich, Barbara. 1978. *For Her Own Good: 150 Years of the Expert's Advice to Women*. Garden City, N.Y.: Anchor Press.

Elbaz, Gilbert. 1990. Measuring AIDS Politicization. Sixth International Conference on AIDS, San Francisco, June 1990.

Elbaz, Gilbert. 1991. *ACT UP: A Crossroads of Social Movements*. Seventh International Conference on AIDS, Florence, 1991.

Elbaz, Gilbert. 1992. *The Sociology of AIDS Activism: The Case of ACT UP/New York, 1987–1992*. Ph.D. dissertation, Sociology Program, City University of New York.

Elias, Norbert. 1971. Sociology of knowledge: new perspectives. *Sociology* 5: 149–168, 355–370.

Epstein, Steve. 1991. Democratic science: AIDS activism and the contested construction of knowledge. *Socialist Review* 21(2): 35–64.

Epstein, Steve. 1996. *Impure Science: AIDS, Activism, and the Politics of Knowledge*. Berkeley: University of California Press.

Erni, John Nguyet. 1994. *Unstable Frontiers: Technomedicine and the Cultural Politics of "Curing" AIDS*. Minneapolis: University of Minnesota Press.

Escobar, Arturo. 1994. Welcome to Cyberia: notes on the anthropology of cyberculture. *Current Anthropology* 35(3): 211–232.

Falcão, Edgard de Cerqueira. 1973. *Oswaldo Cruz Monumenta Historica*. São Paulo: Brasiliensia Documenta VI.

Farmer, Paul. 1992. *AIDS and Accusation: Haiti and the Geography of Blame*. Berkeley: University of California Press.

Farmer, Paul. 1994. AIDS talk and the constitution of cultural models. *Soc Sci Med* 38(6): 801–809.

Farmer, Paul, Shirley Lindenbaum, and Mary-Jo DelVecchio Good. 1993. Women, poverty and AIDS: an introduction. *Culture, Medicine, and Psychiatry* 17: 387–397.

Fausto-Sterling, Anne. 1992 [1985]. *Myths of Gender*. New York: Basic Books.

Featherstone, Mike, ed. 1990. *Global Culture: Nationalism, Globalization and Modernity*. London: Sage.

Featherstone, Mike, Scott Lash, and Roland Robertson, eds. 1995. *Global Modernities*. London: Sage.

Fee, Elizabeth, and Daniel Fox, eds. 1988. *AIDS: The Burdens of History*. Berkeley: University of California Press.

Fee, Elizabeth, and Daniel Fox, eds. 1992. *AIDS: The Making of a Chronic Disease*. Berkeley: University of California Press.

Feinberg, David. 1991. *Spontaneous Combustion*. New York: Viking.

Feinberg, David. 1995. *Queer and Loathing: Rants and Raves of a Raging AIDS Clone*. New York: Viking/Penguin.

Feldman, Douglas A., ed. 1994. *Global AIDS Policy*. Westport, Conn.: Bergin and Garvey.

Feldman, Douglas A., and Thomas M. Johnson, eds. 1986. *The Social Dimensions of AIDS*. New York: Praeger.

Fernandes, Ana Maria. 1987. The Scientific Community and the State in Brazil: The Role of the Brazilian Society for the Advancement of Science, 1948–1980. Ph.D. dissertation, Oxford University.

Fernandes, Florestan. 1979. The Negro in Brazilian Society: Twenty-five Years Later. In *Brazil: Anthropological Perspectives,* ed. Maxine Margolis and William Carter, 96–113. New York: Columbia University Press.

Fernandes, João Claudio Lara. 1994. AIDS e favelas: O programa regionalizado de controle de AIDS. *Boletim Abia* 23: 11.

Fernandes, Rubem César. 1988. Sem fins lucrativos. In *Sem fins lucrativos*. Leilah Landim, ed. 8–23. Rio de Janeiro: ISER.

Fernandes, Rubem César, and Leandro Piquet Carneiro. 1995. Brazilian NGOs in the 1990s: A Survey. In *New Paths to Democratic Development in Latin America: The Rise of NGO–Municipal Collaboration,* ed. Charles Reily, 71–84. Boulder, Colo.: Lynne Rienner.

Ferri, Mário Guimarães, e Shozo Motoyama. 1979. *História das Ciências no Brasil*. São Paulo: EDUSP.

Feyerabend, Paul. 1993. *Against Method.* 3rd ed. New York: Verso.

Filgueiras, Ana. 1994. Out-of-school youth: a need for NGO and governmental collaboration. *Exchanges* 1994(1): 1–4.

Fleck, Ludwik. 1979. *Genesis and Development of a Scientific Fact.* Chicago: University of Chicago Press.

Fleming, Alan F., Manuel Carballo, David W. FitzSimons, Michael R. Bailey, and Jonathan Mann, eds. 1988. *The Global Impact of AIDS*. New York: Alan R. Liss.

Forsythe, Diana. 1993. Engineering knowledge: the social construction of knowledge in artificial intelligence. *Social Studies of Science* 23(3): 445–477.

Foucault, Michel. 1969. *L'archéologie du Savoir*. Paris: Gallimard.

Fraga Filho, Clementino. 1990. *A Implantação do Hospital Universitário da UFRJ (1974/1978)*. Rio de Janeiro: Fundação Universitária José Bonifácio.

Frank, Andrew Gunder. 1967. *Capitalism and Underdevelopment in Latin America: Historical Studies in Chile and Brazil*. New York: Monthly Review Press.

Frankenberg, Ronald. 1994. The impact of HIV/AIDS on concepts relating to risk and culture within British community epidemiology: candidates or targets for prevention? *Soc Sci Med* 38(10): 1325–1335.

Franklin, Sarah. 1995a. Postmodern Procreation: a Cultural Account of Assisted Reproduction. In *Conceiving the New World Order,* ed. Faye Ginsburg and Rayna Rapp, 323–345. Berkeley: University of California Press.

Franklin, Sarah. 1995b. Science as culture, cultures as science. *Annual Review of Anthropology* 24: 163–184.

Frazer, James George. 1900. *The Golden Bough: A Study in Magic and Religion.* 2nd ed. London: Macmillan.

Fredrickson, Donald S. 1977. Health and the Search for New Knowledge. In *Doing Better and Feeling Worse: Health in the United States,* ed. John Knowles, 159–170. New York: Norton.

Freudenthal, Gad, and Ilana Löwy. 1988. Ludwick Fleck's roles in society: a case-study using Joseph Ben-David's paradigm for a sociology of knowledge. *Social Studies of Science* 1894: 625–651.

Friedman, Jonathan. 1994. *Cultural Identity and Global Process*. London: Sage.

Friedman, Samuel R., Don C. Des Jarlais, and Douglas S. Goldsmith. 1989. An overview of current AIDS prevention efforts aimed at intravenous drug users. *Journal of Drug Issues* 19(1): 93–112.

Fry, Peter. 1982. *Para Inglês Ver.* Rio de Janeiro: Zahar.

Fry, Peter. 1989. Prefácio. In *A Construção da Igualdade,* by Edward MacRae. 11–15. Campinas: Editora da Unicamp.

Fry, Peter. 1995. Male Homosexuality and Afro-Brazilian Possession Cults. In *Latin American Male Homosexualities,* ed. Stephen O. Murray, 193–220. Albuquerque: University of New Mexico Press.

Fry, Peter, and Edward MacRae. 1985. *O que é a Homossexualidade*. São Paulo: Abril Cultural/Brasiliense.

Fujimura, Joan. 1987. Constructing "do-able" problems in cancer research: articulating alignment. *Social Studies of Science* 17(2): 257–293.

Fujimura, Joan. 1988. The molecular biology bandwagon in cancer research. *Social Problems* 35(3): 261–283.

Fujimura, Joan, and Danny Chou. 1994. Dissent in science: styles of scientific practice and the controversy over the cause of AIDS. *Soc Sci Med* 39(8): 1017–1036.

Fuller, Steve. 1993. *Philosophy, Rhetoric, and the End of Knowledge: The Coming of Age of Science and Technology Studies*. Madison: University of Wisconsin Press.

Gaillard, Jacques. 1991. *Chercheurs des Pays en Developpement. Scientists in the Third World.* Lexington: University Press of Kentucky.

Gallo, Robert. 1991. *Virus Hunting: AIDS, Cancer, and the Human Retrovirus—A Story of Scientific Discovery.* New York: Basic Books.

Gallo, Robert, and Luc Montagnier. 1988. AIDS in 1988. *Sci Am* 259(4) (special issue): 40–48.

Galvão, Jane. 1992a. *AIDS e imprensa: Um estudo de antropologia social.* M.A. dissertation. Programa de Pós-Graduação em Antropologia Social, Museu Nacional, UFRJ, Rio de Janeiro.

Galvão, Jane. 1992b. AIDS e ativismo: O surgimento e a construção de novas formas de solidariedade. Paper presented to the seminar AIDS e Ativismo Social e Político, IMS/UERJ, Rio de Janeiro.

Gamson, Joshua. 1991. Silence, Death, and the Invisible Enemy: AIDS Activism and the Social Movement "Newness." In *Ethnography Unbound,* ed. Michael Burawoy, 5–57. Berkeley: University of California Press.

Garrett, Laurie. 1992. The Next Epidemic. In *AIDS in the World,* ed. Mann et al., 825–843. Cambridge, Mass.: Harvard University Press.

Garrett, Laurie. 1994. *The Coming Plague*. New York: Farrar, Straus and Giroux.

Gaspar, Júlio Dias. 1992. Cadernos Pela VIDDA: a resposta de um grupo ativista para enfrentar a epidemia de AIDS. *Boletim ABIA* 2 (boletim especial): 2–4.

Gay Men's Health Crisis. 1995. *GMHC Facts, October 1995*. New York: GMHC.

Gilbert, G. Nigel, and Michael Mulkay. 1984. *Opening Pandora's Box: A Sociological Analysis of Scientists Discourse*. Cambridge: Cambridge University Press.

Gilman, Sander. 1988. *Disease and Representation: Images of Illness from Madness to AIDS*. Ithaca, N.Y.: Cornell University Press.

Ginsburg, Faye D., and Rayna Rapp, eds. 1995. *Conceiving the New World Order: The Global Politics of Reproduction*. Berkeley: University of California Press.

Global AIDS Policies Coalition. 1991. Coalizão Global das Políticas Contra a AIDS. *Boletim Abia* 15: 4–5.

Goldin, Carol S. 1994. Stigmatization and AIDS: critical issues in public health. *Soc Sci Med* 39(9): 1359–1366.

Goldsmith, Douglas and Samuel Friedman. 1991. La droge, le sexe, le SIDA, et la survie dans la rue. Les voix de cinq femmes. *L'univers du Sida. Anthropologie et Société* 15(2–3) (special issue): 13–36.

Goldstein, Donna M. 1994. AIDS and women in Brazil: the emerging problem. *Soc Sci Med* 39(7): 919–929.

Gomez, Marianne. 1992. Plus jamais sans les malades. *Autrement/Mutations* 130: 40–49.

Good, Byron. 1994. Medicine, *Rationality and Experience: An Anthropological Perspective*. Cambridge: Cambridge University Press.

Good, Charles M. 1995. Incentives to lower the incidence of HIV/AIDS in Africa. *Soc Sci Med* 40(4): 419–424.

Good, Mary-Jo DelVecchio. 1995. Cultural studies of biomedicine: an agenda for research. *Soc Sci Med* 41(4): 461–475.

Goonatilake, Susantha. 1993. Modern Science and the Periphery: The Characteristics of Dependent Knowledge. In *The "Racial" Economy of Science,* ed. Sandra Harding, 259–267. Bloomington: Indiana University Press.

Gorman, Michael. 1986. The AIDS Epidemic in San Francisco: Epidemiological and Anthropological Perspectives. In *Anthropology and Epidemiology,* ed. Craig R. Janes, Ron Stall, and Sandra M. Gifford, 157–172. Dordrecht: D. Reidel.

Gorman, Michael. 1991. A special window: anthropological reflections on the HIV epidemic among gay men. *The Journal of Sex Research* 28(2): 263–273.

Gould, Peter. 1993. *The Slow Plague: A Geography of the AIDS Pandemic*. Cambridge, Mass.: Blackwell.

Granjeiro, Alexandre. 1994. O Perfil Socioeconômico dos Casos de AIDS da Cidade de São Paulo. In *A AIDS no Brasil,* ed. Parker et al., 91–125. Rio de Janeiro: Relume–Dumará.

Graubard, Stephen, ed. 1990. *Living with AIDS*. Cambridge, Mass.: MIT Press.

Green, Jesse. 1995. Who Put the Lid on gp120? *New York Times Magazine*, March 26: 1, 50–57, 74, 82.

Greenberg, David F. 1988. *The Construction of Homosexuality.* Chicago: University of Chicago Press.

Greenwood, Davydd, Shirley Lindenbaum, Margaret Lock, and Allan Young, eds. 1988. Medical anthropology. *American Ethnologist* 15(1) (special issue).

Gross, Paul R., and Norman Levitt. 1994. *Higher Superstition: The Academic Left and Its Quarrels with Science*. Baltimore: Johns Hopkins University Press.

Grupo de Apoio à Prevenção da AIDS, Rio de Janeiro. 1989. A vida continua. *Boletim Abia* 6: 2.

Grupo Gay da Bahia. 1990. Sugestões do Grupo Gay da Bahia à Rede Brasileira de Grupos de Prevenção à AIDS. Unpublished article.

Grupo Pela VIDDA RJ, Grupo Pela VIDDA SP and ABIA. 1992. Dossiê 1: Vacinas.
Guedes, Regina, and Alvaro Matida. 1989. Entrevista com o dr. Alvaro Matida e a dra. Regina Guedes, do departamento de vigilância epidemiológica da secretaria estadual de saúde/RJ. *Boletim Abia* 6: 6–8.

Guibert, Hervé. 1990. *A l'ami qui ne m'a pas sauvé la vie*. Paris: Gallimard.
Guibert, Hervé. 1991. *Le protocole compassional*. Paris: Gallimard.
Guibert, Hervé. 1992. *Cytomégalovirus: Journal d'une Hospitalisation*. Paris: Editions Du Seuil.
Guimarães, Carmen Dora. 1977. *O Homossexual Visto por Entendidos*. M.A. dissertation. Programa de Pós-Graduação em Antropologia Social, Museu Nacional, UFRJ, Rio de Janeiro.
Guimarães, Carmen Dora, Herbert Daniel, and Jane Galvão. 1988. A questão do preconceito. *Boletim Abia* 3: 2–3.
Gupta, Geeta Rao and Ellen Weiss. 1993. Women's lives and sex: implications for AIDS prevention. *Culture Medicine and Psychiatry* 17(4): 399–412.
Gurgel, Artur Amaral. 1988. O GAPA está precisando de uma sede (Entrevista). *Boletim Abia* 2: 4–5.
Habermas, Jurgen. 1981. Modernity versus postmodernity. *New German Critique* 22: 3–14.
Habermas, Jurgen. 1983. The Incomplete Project of Modernity. In *Anti-Aesthetic: Essays in Postmodern Culture,* ed. Hal Foster, 3–15. Port Townsend, Wash.: Bay Press.
Hahn, Robert A. 1991. What should behavioral scientists be doing about AIDS? *Soc Sci Med* 33(1): 1–3.
Hahn, Robert, and Atwood Gaines, eds. 1985. *Physicians of Western Medicine: Anthropological Approaches to Theory and Practice*. Dordrecht: D. Reidel.
Halperin, David M. *1995. Saint = Foucault? Towards a Gay Hagiography*. New York: Oxford University Press.
Hannerz, Ulf. 1988. *Culture between Center and Periphery: Toward a Macroanthropology.* Vega Day Symposium on Culture in Complex Societies. Stockholm: Swedish Society for Anthropology and Geography.
Hannerz, Ulf. 1992. *Cultural Complexity: Studies in the Social Organization of Meaning.* New York: Columbia University Press.
Haraway, Donna. 1989. The biopolitics of postmodern bodies: determinations of self in immune system discourse. Differences: *A Journal of Feminist Cultural Studies* 1(1): 3–43.
Haraway, Donna. 1991. *Simians, Cyborgs, and Women: The Reinvention of Nature*. New York: Routledge.
Haraway, Donna. 1995. Situated Knowledges: The Science Question in Feminism and the Privilege of Partial Perspective. In *Technology and the Politics of Knowledge,* ed. Andrew Feenberg and Alastair Hannay, 175–194. Bloomington: Indiana University Press.
Harden, Victoria A., and Dennis Rodrigues. 1993. Context for a New Disease: Aspects of Biomedical Research Policy in the United States before AIDS. In *AIDS and Contemporary History,* ed. Virginia Berridge and Philip Strong, 182–202. Cambridge: Cambridge University Press.
Harding, Sandra. 1986. *The Science Question in Feminism*. Ithaca, N.Y.: Cornell University Press.
Harding, Sandra. 1998. *Is Science Multicultural?* Bloomington: Indiana University Press.
Harding, Sandra, ed. 1993. *The "Racial" Economy of Science: Toward a Democratic Future.* Bloomington: Indiana University Press.
Harding, Sandra, and Jean F. O'Barr, eds. 1987. *Sex and Scientific Inquiry*. Chicago: University of Chicago Press.

Harrington, Mark. 1992. Draft of a Speech on Pathogenesis and Activism. Eighth International Conference on AIDS, Amsterdam, July 21.

Harrington, Mark. 1994. The Community Research Initiative (CRI) of New York: Clinical Research and Prevention Treatments. In *AIDS Prevention Services: Community Based Research,* ed. Johannes P. Van Vugt, 179–198. Westport, Conn.: Bergin and Garvey.

Harwood, Jonathan. 1986. Ludwick Fleck and the sociology of knowledge. *Social Studies of Science* 16(1): 173–187.

Heath, Diana, and Paul Rabinow, eds. 1993. Bio-Politics: The anthropology of the new genetics and immunology. *Culture Medicine and Psychiatry* 17 (special issue).

Heise, Lori L., and Christopher Elias. 1995. Transforming AIDS prevention to meet women's needs: a focus on developing countries. *Soc Sci Med* 40(7): 931–943.

Henderson, Sheila. 1991. Care: What's in It for Her? In *AIDS: Responses, Interventions and Care,* ed. Peter Aggleton, Graham Hart, and Peter Davies, 261–272. London: Falmer Press.

Herdt, Gilbert, et al. 1990. AIDS on the planet: the plural voices of anthropology. *Anthropology Today* 6(3) (special issue): 10–15.

Herdt, Gilbert, William Leap, and Melanie Sovine, eds. 1991. Anthropology, AIDS and sex. *The Journal of Sex Research* 28(2).

Herdt, Gilbert, and Shirley Lindenbaum. 1992. *The Time of AIDS: Social Analysis, Theory, and Method.* London: Sage.

Héritier-Augé, Françoise. 1992. Ce mal invisible et sournois. *Autrement/Mutations* 130: 148–157.

Hess, David. 1992. Introduction: The New Ethnography and the Anthropology of Science and Technology. In *The Anthropology of Science and Technology,* ed. David Hess and Linda Layne. *Knowledge and Society*, 9: 1–26 (special issue).

Hess, David, and Linda Layne, eds. 1995. *Science and Technology in a Multicultural World: The Cultural Politics of Facts and Artifacts*. New York: Columbia University Press.

Hessen, Boris. 1931. The Social and Economic Roots of Newton's Principia. In *Science at the Crossroads,* ed. Bukharin et al. London: Frank Cass.

Hicks, Diana, and Jonathan Potter. 1991. Sociology of scientific knowledge: a reflexive citation analysis or science disciplines and disciplining science. *Social Studies of Science* 21: 459–501.

Hilgartner, Stephen. 1995. The Human Genome Project. In *Handbook of Science and Technology Studies,* ed. Jasanoff et al., 302–315. London: SAGE.

Horton, Meurig. 1989. Bugs, Drugs and Placebos: The Opulence of Truth; or, How to Make a Treatment Decision in an Epidemic. In *Taking Liberties,* ed. Erica Carter and Simon Watney, 161–181. London: Serpent's Tail.

Horton, Meurig. 1993. The Role of Homosexuals in the Fight against AIDS. Speech to the Plenary Assembly, Ninth International Conference on AIDS, Berlin, June.

Horton, Meyrick, with Peter Aggleton. 1989. Perverts, Inverts, and Experts: The Cultural Production of an AIDS Research Paradigm. In *AIDS: Social Representations, Social Practice,* ed. Peter Aggleton, Graham Hart, and Peter Davies, 74–100. London: Falmer Press.

Hospital Evandro Chagas. 1990. Boletim Epidemiológico do Hospital Evandro Chagas, FIOCRUZ. III(2).

Ianni, Octavio. 1992. *A Sociedade Global*. Rio de Janeiro: Civilização Brasileira.
Inhorn, Marcia, and Peter J. Brown. 1990. The anthropology of infectious disease. *Annual Review of Anthropology* 19.
Instituto Superior de Estudos de Religião. 1985. Um vírus só não faz a doença. *Comunicações do ISER* 17.
Instituto Superior de Estudos de Religião. 1988. AIDS: Somos todos mortais. *Comunicações do ISER* 31 (special issue).
Instituto Superior de Estudos de Religião. 1990. *Fala, Mulher da Rua*. Rio de Janeiro: ISER.
Jacobus, Mary, Evelyn Fox Keller, and Sally Shuttleworth, eds. 1990. *Body/Politics: Women and the Discourse of Sciences*. New York: Routledge.
Jaret, Peter. 1986. The Immune System: The Wars Within. *National Geographic*.
Jasanoff, Sheila, Gerald E. Markle, James C. Petersen, and Trevor Pinch, eds. 1995. *Handbook of Science and Technology Studies*. London: Sage.
Kalichman, Artur O. 1994. Pauperização e banalização de uma epidemia. In *A Epidemiologia Social da AIDS*, ed. Richard Parker and Jane Galvão, 20–26. Rio de Janeiro: ABIA–IMS.
Kamminga, Harmke. 1994. The harmonisation of Elie Metchnikoff: making sense of cellular immunity. *Studies in History and Philosophy of Science* 25(1): 131–145.
Kane, Stephanie. 1991. HIV, heroin and heterosexual relations. *Soc Sci Med* 32(9): 1037–1050.
Kaplan, Howard B. 1991. Social psychology of the immune system: a conceptual framework and review of the literature. *Soc Sci Med* 33(8): 909–923.
Keller, Evelyn Fox. 1983. *A Feeling for the Organism: The Life and Work of Barbara McClintock*. New York: W. H. Freeman and Company.
Keller, Evelyn Fox. 1985. *Reflections on Gender and Science*. New Haven, Conn.: Yale University Press.
Keller, Evelyn Fox. 1995. *Refiguring Life: Metaphors of Twentieth-Century Biology*. New York: Columbia University Press.
King, A. B., ed. 1991. *Culture, Globalization and the World System: Contemporary Conditions for the Representation of Identity*. Binghamton: State University of New York, and London: Macmillan.
Kinsella, James. 1989. *Covering the Plague: AIDS in the American Media*. New Brunswick, N.J.: Rutgers University Press.
Kinsey, Alfred C., Wardell Pomeroy, and Clyde E. Martin. 1948. *Sexual Behavior in the Human Male*. Philadelphia: W. B. Saunders.
Kinsey, Alfred, et al. Institute for Sex Research. 1953. *Sexual Behavior in the Human Female*. Philadelphia: W. B. Saunders.
KIT and SAfAIDS. 1995. Experiences of NGO networks working with AIDS. *Exchanges* 1995(1): 10–11.
Knorr-Cetina, Karin. 1981. *The Manufacture of Knowledge: An Essay in the Constructivist and Contextual Nature of Science*. Oxford: Pergamon Press.
Knorr-Cetina, Karin, and Michael Mulkay, eds. 1983. *Science Observed: Perspectives on the Social Study of Science*. London: Sage.
Knowles, John, ed. 1977. *Doing Better and Feeling Worse: Health in the United States*. New York: Norton.
Koester, Stephen. 1994. Applying Ethnography to AIDS Prevention among IV Drug Users and Social Policy Implications. In *AIDS Prevention Services: Community*

Based Research, ed. Johannes P. Van Vugt, 35–57. Westport, Conn.: Bergin and Garvey.

Kramer, Larry. 1978. *Faggots*. New York: Random House.

Kramer, Larry. 1985. *The Normal Heart*. New York: New American Library.

Kramer, Larry. 1989. *Report from the Holocaust: The Making of an AIDS Activist*. New York: Saint Martin's Press.

Kritsky, Afrânio. n.d. Interview given to Cristina Câmara (transcript). Rio de Janeiro: Pela VIDDA Group.

Kruif, Paul de. 1926. *Microbe Hunters*. New York: Harvest/HBJ.

Kuhn, Thomas S. 1962. *The Structure of Scientific Revolutions*. Chicago: University of Chicago Press.

Kulick, Henrika. 1983. The sociology of knowledge: retrospect and prospect. *Annual Review of Sociology* 9: 287–310.

Labinger, Jay A. 1995. Science as culture: a view from the petri dish. *Social Studies of Science* 25: 285–306. With commentary.

Lancaster, Roger. 1994. *Life Is Hard: Machismo, Danger, and the Intimacy of Power in Nicaragua*. Berkeley: University of California Press.

Landim, Leilah, ed. 1988. *Sem fins lucrativos*. Rio de Janeiro: ISER.

Lapierre, Dominique. 1990. *Plus grand que l'amour*. Paris: Robert Laffont.

Larvie, S. Patrick. 1995. Self-Help and Bad Self-Esteem: Behavioral Interventions to Prevent AIDS and Emergent Theories of Sexual Citizenship. Paper presented at the Annual Meeting of the American Anthropological Association, Washington, D.C.

Latour, Bruno. 1983. Comment redistribuer le grand partage? *Revue de Synthèse* 110: 203–236.

Latour, Bruno. 1984. *Les Microbes: Guerre et Paix, suivi d'Irréductions*. Paris: Métailé.

Latour, Bruno. 1987. *Science in Action: How to Follow Scientists and Engineers through Society*. Cambridge, Mass.: Harvard University Press.

Latour, Bruno. 1990. Postmodern? No, simply amodern!: steps towards an anthropology of science. *Studies in the History and Philosophy of Science*.

Latour, Bruno, and Steve Woolgar. 1979. *Laboratory Life: The Construction of Scientific Facts*. Introduction by Jonas Salk. Princeton, N.J.: Princeton University Press.

Laudan, Larry. 1990. *Science and Relativism: Some Key Controversies in the Philosophy of Science*. Chicago: University of Chicago Press.

Laurell, A. Cristina. 1976. Algunos problemas teóricos y conceptuales de la epidemiología social (mimeo).

Laurell, A. Cristina. 1987. *Proceso de Producción y Salud: una Proposta Teórico Metodológica y Técnica y su Utilización en un Estudio de Caso*. Mexico City: UNAM. Ph.D. dissertation.

Lauritsen, John. 1990. *Poison by Prescription: The AZT Story*. New York: Asklepios.

Leslie, Charles. 1990. Scientific racism: reflections on peer review, science and ideology. *Soc Sci Med* 31(8): 891–912.

Lévi-Strauss, Claude. 1962a. *La Pensée Sauvage*. Paris: Plon.

Lévi-Strauss, Claude. 1962b. *Le Totémisme Aujourd'hui*. Paris: Presses Universitaires de France.

Levidow, Les, ed. 1986. *Science as Politics*. London: Free Association Books.

Lewis, Diane K., and John K. Watters. 1989. Human immunodeficiency virus seroprevalence in female intravenous drug users: the puzzle of black women's risk. *Soc Sci Med* 29(9): 1071–1076.

Lewontin, R. C. 1992. *Biology as Ideology: The Doctrine of DNA*. New York: HarperCollins.

Lindenbaum, Shirley. 1979. *Kuru Sorcery: Disease and Danger in the New Guinea Highlands*. Mountain View, Calif.: Mayfield.

Lindenbaum, Shirley. 1992. Knowledge and Action in the Shadow of AIDS. In *The Time of AIDS*, ed. Gilberdt Herdt and Shirley Lindenbaum, 319–334. London: Sage.

Lindenbaum, Shirley, and Margaret Lock, eds. 1993. *Knowledge, Power and Practice: The Anthropology of Medicine and Everyday Life*. Berkeley: University of California Press.

Lindenbaum, Shirley, et al. 1993. The HIV/AIDS Epidemic in New York City. In *The Social Impact of AIDS in the United States*, ed. Albert Jonsen and Jeff Stryker, 201–242. Washington, D.C.: National Academy Press.

Lock, Margaret, and Deborah Gordon, eds. 1988. *Biomedicine Examined*. London: Kluwer Academic Press.

Longino, Helen A. 1995. Knowledge, Bodies and Values: Reproductive Technologies and Their Scientific Contexts. In *Technology and the Politics of Knowledge*, ed. Andrew Feenberg and Alastair Hannay, 195–210. Bloomington: Indiana University Press.

Lopes, José Leite. 1987. *Ciência e Desenvolvimento (Ensaios). Segunda Edição Revista e Aumentada*. Rio de Janeiro: Tempo Brasileiro.

Löwy, Ilana. n.d. Les faits scientifiques et leur public: l'historie de la réaction de Wasserman pour la détection de la syphilis. Unpublished manuscript.

Löwy, Ilana. 1989. Biomedical research and the constraints of medical practice: James Bumgarder Murphy and the early discovery of the role of lymphocytes in immune reactions. *Bull Hist Med* 63: 356–391.

Löwy, Ilana. 1990. Variances in meaning in discovery accounts: the case of contemporary biology. *Historical Studies in the Physical and Biological Sciences* 21(1): 87–121.

Löwy, Ilana. 1991. The Immunological Construction of the Self. In *Organism and the Origins of Self*, ed. Alfred Tauber, 43–75. Dordrecht: Kluwer Academic Publishers.

Löwy, Ilana. 1992. The strength of loose concepts—boundary concepts, federative experimental strategies and disciplinary growth: the case of immunology. *History of Science* 30: 371–396.

Löwy, Ilana. 1995. On hybridizations, networks and new disciplines: the Pasteur Institute and the development of microbiology in France. *Studies in History and Philosophy of Science* 25(5): 655–688.

Loyola, Maria Andréa, ed. 1994. *Aids e Sexualidade: O Ponto de Vista das Ciências Humanas*. Rio de Janeiro: Relume–Dumará/UERJ.

Lynch, Michael. 1985. *Art and Artifact in Laboratory Science: A Study of Shop Work and Shop Talk in a Research Laboratory*. London: Routledge and Kegan Paul.

Ma, Pearl, and Donald Armstrong, eds. 1984. *The Acquired Immune Deficiency Syndrome and Infections of Homosexual Men*. New York: Yorke Medical Books.

MacGarrahan, Peggy. 1992. Transformation and Transcendence: Caring for HIV Infected Patients in New York City. Ph.D. dissertation, Anthropology Program, City University of New York.

MacKenzie, Nancy F., ed. 1991. *The AIDS Reader: Social, Political, Ethical Issues*. New York: Meridien.

MacKeown, T. 1976. *The Modern Rise of Population.* New York: Academic Press.

MacNeil, Maureen, and Sarah Franklin. 1991. Science and Technology: Questions for Feminism and Cultural Studies. In *Off-Centre: Feminism and Cultural Studies,* ed. Sarah Franklin, Celia Lury, and Jackie Stacey, 129–146. London: HarperCollins.

MacNeill, William. 1976. *Plagues and Peoples.* New York: Doubleday.

MacRae, Edward. 1990. *A Construção da Igualdade: Identidade Sexual e Política no Brasil da "Abertura."* Campinas: Editora da Unicamp.

Magaña, J. Raúl. 1991. Sex, Drugs and HIV: An ethnographic approach. *Soc Sci Med* 33(1): 5–9.

Malinowski, Bronislaw. 1948. *Magic, Science and Religion.* Glencoe, Ill.: The Free Press.

Mann, Jonathan. 1990. Carta do Dr. Mann ao II encontro da rede Brasileira de solidariedade. *Boletim Abia* 9: 11.

Mann, Jonathan. 1991. HIV/AIDS Prevention and Care in the 1990s: The Pandemic, Critical Issues and Gaps. Report to the Ford Foundations, attachment 3.

Mann, Jonathan, James Chin, Peter Piot, and Thomas Quinn. 1988. The international epidemiology of AIDS. *Sci Am* 259(4): 82–89.

Mann, Jonathan, Daniel J. M. Tarantola, and Thomas W. Netter, eds. 1992. *AIDS in the World: A Global Report.* Cambridge, Mass.: Harvard University Press.

Mann, Jonathan, Daniel J. M. Tarantola, and Thomas W. Netter, eds. 1993. *A AIDS no Mundo.* Rio de Janeiro: ABIA/IMS–UERJ/Relume–Dumará.

Mannheim, Karl. 1936. *Ideology and Utopia: An Introduction to the Sociology of Knowledge.* New York: Harvest/HBJ.

Marcus, George E., ed. 1995. *Technoscientific Imaginaries: Conversations, Profiles, and Memoirs.* Chicago: University of Chicago Press.

Marcus, George E., and Michael M.J. Fischer. 1986. *Anthropology as Cultural Critique: An Experimental Moment in the Human Sciences.* Chicago: University of Chicago Press.

Marques, Marília Bernardo. 1989. *Limites ao Desenvolvimento Scientífico e Tecnológico em Saúde no Brasil. Série Política de Saúde, 9.* Rio de Janeiro: Fiocruz.

Marques, Marília Bernardo. 1993. Transferência de tecnologia em AIDS: conflitos e oportunidades. *Boletim Abia* 18: 3–4.

Martin, Emily. 1987. *The Woman in the Body: A Cultural Analysis of Reproduction.* Boston: Beacon Press.

Martin, Emily. 1990. Toward an anthropology of immunology: the body as a nation-state. *Medical Anthropology Quarterly* 4(4): 410–426.

Martin, Emily. 1992. The end of the body? *American Ethnologist* 19: 121–140.

Martin, Emily. 1993. Histories of immune systems. *Culture, Medicine and Psychiatry* 17: 67–76.

Martin, Emily. 1994. *Flexible Bodies: Tracking Immunity and Infection in American Culture from the Days of Polio to the Age of AIDS.* Boston: Beacon Press.

Marx, Karl. 1973. *Grundrisse: Foundations of the Critique of Political Economy.* New York: Random House.

Mass, Lawrence. 1990a. *Dialogues of the Sexual Revolution I: Homosexuality and Sexuality.* New York: Harrington Park Press.

Mass, Lawrence. 1990b. *Dialogues of the Sexual Revolution II: Homosexuality as Behavior and Identity.* New York: Harrington Park Press.

Mendelsohn, Everett, and Yehuda Elkana, eds. 1981. *Sciences and Cultures: Anthropological and Historical Studies of the Sciences.* Dordrecht: D. Reidel.

Merton, Robert. 1973. *The Sociology of Science.* Chicago: University of Chicago Press.

Minayo, Maria Cecília de Souza, ed. 1995. *Os Muitos Brasis: Saúde e população na década de 80*. São Paulo–Rio de Janeiro: HUCITEC–ABRASCO.

Ministério da Saúde. 1987. Recomendações para prevenção e controle da infecção pelo vírus HIV (SIDA–AIDS). Brasília: Programa Nacional de Controle de Doenças Sexualmente Transmissíveis e AIDS.

Ministério da Saúde. 1988a. SIDA/AIDS: Manual de Condutas Clínicas. Brasília: Divisão Nacional de Doenças Sexualmente Transmissíveis—SIDA/AIDS.

Ministério da Saúde. 1988b. Aspectos clínicos, laboratoriais e terapêuticos das doenças sexualmente transmissíveis: Manual do aluno. Brasília: Divisão Nacional de Controle de Doenças Sexualmente Transmissíveis—SIDA/AIDS.

Ministério da Saúde. 1994a. Catálogo de organizações não-governamentais. With partial English version. Brasília: Ministério da Saúde, Secretaria de Assistência à Saúde, Programa Nacional de Doenças Sexualmente Transmissíveis/AIDS.

Ministério da Saúde. 1994b. Abstracts from Brazil. Tenth International Conference on AIDS/International Conference on SDT, Yokohama, August 7–12, 1994. Brasília: Ministério da Saúde, Secretaria de Assistência à Saúde, Programa Nacional de Doenças Sexualmente Transmissíveis/AIDS.

Mogadham, A. A. 1991. *The North–South Science and Technology Gap*. New York: Garland Publishing.

Mondragón, Delf, Bradford Kirkman-Liff, and Eugene S. Schneller. 1991. Hostility to people with AIDS: risk perception and demographic factors. *Soc Sci Med* 32(10): 1137–1142.

Monette, Paul. 1988. *Borrowed Time: An AIDS Memoir*. San Diego: Harcourt, Brace and Jovanovich.

Monteiro, Simone, C. Castello Branco, J. Galvão, and R. Parker. 1994. AIDS prevention in schools through partnership between governments and NGOs. Poster abstract Pd0599 presented at the Tenth International Conference on AIDS, Yokohama, Abstract Book 2: 355.

Moraes, Claudia, and Sérgio Carrara. 1985a. AIDS: Um vírus só não faz a doença. *Comunicações do ISER* 17: 5–19.

Moraes, Claudia, and Sérgio Carrara. 1985b. Um mal de folhetim. *Comunicações do ISER* 17: 20–31.

Morazé, Charles. 1979. *Science and the Factors of Inequality*. Paris: UNESCO.

Morazé, Charles, et al. 1980. Le point critique. Studies prepared under the direction of Charles Morazé: reports for the U.N. Conference in science and technology for development, Vienna, 1979, prepared by Gerard Guillet. Paris: Presses Universitaires de France.

Moreira, Marcos. 1974. *Oswaldo Cruz*. São Paulo: Editora Três.

Morel, Regina Lúcia de Moraes. 1979. *Ciência e Estado: A política científica do Brasil*. São Paulo: T.A. Queiroz.

Morris, R. J. 1976. *Cholera 1832: The Social Responses to the Epidemic*. London: Croom Helm.

Morse, Stephen S. 1991. Emerging viruses: defining the rules for viral traffic. *Perspect Biol Med* 34(3): 387–409.

Mota, Murillo Peixoto da. 1994. *AIDS: Expressão de Complexidade*. Rio de Janeiro: mim.

Motoyama, Shozo. 1988. História da Ciência no Brasil. Apontamento para uma análise crítica. *Quipu* 5(2): 167–189.

Mott, Luiz. 1985. *Relações raciais entre homossexuais no Brasil Colônia. Produção e Trangressões, Revista Brasileira de História 1.0* São Paulo: Marco Zero.

Mott, Luiz. 1988a. *Escravidão, Homosexualidade e Demonologia*. São Paulo: Icone.
Mott, Luiz. 1988b. *O Sexo Proibido: Virgens, Gays e Escravos nas Garras da Inquisição*. São Paulo: Papirus.
Mott, Luiz. 1988c. Correspondência. *Boletim Abia* 3: 6.
Mott, Luiz. 1995. The Gay Movement and Human Rights in Brazil. In *Latin American Male Homosexualities*, ed. Stephen O. Murray, 221–230. Albuquerque: University of New Mexico Press.
Mulkay, Michael. 1972. *The Social Process of Innovation: A Study in the Sociology of Science*. London: Macmillan.
Mulkay, Michael. 1979. *Science and the Sociology of Knowledge*. London: Allen & Unwin.
Murray, Stephen O., ed. 1995. *Latin American Male Homosexualities*. Albuquerque: University of New Mexico Press.
Murray, Stephen O., and Kenneth W. Payne. 1989. The Social Classification of AIDS in American Epidemiology. In *The AIDS Pandemic: A Global Emergency*, ed. Ralph Bolton, 23–36. New York: Gordon and Breach.
Mussauer de Lima, Ronaldo. 1994. Promoting the dignity and rights of people living with HIV/AIDS. *Exchanges* 1994(3): 4–6.
National Health Education Committee. 1963. *What Are the Facts about World Health and the World Health Organization?* New York: National Health Education Committee.
National Research Council. 1989. *AIDS: Sexual Behavior and Intravenous Drug Use*. Washington, D.C.: National Academy Press.
Navarro, Vicente, ed. 1981. *Imperialism, Health and Medicine*. Farmingdale, N.Y.: Baywood Publishing.
Neaigus, Alan. 1994. The relevance of drug injectors networks for understanding and preventing HIV infection. *Soc Sci Med* 38(1): 67–78.
Neagius, Alan, et al. 1990. Effects of outreach intervention on risk reduction among intravenous drug users. *AIDS Education and Prevention* 2(4): 253–271.
Nelkin, Dorothy, David P. Willis, and Scott V. Parris. 1991. *A Disease of Society: Cultural and Institutional Responses to AIDS*. New York: Cambridge University Press.
New York Times. 1989. AZT's inhuman cost. *New York Times*, August 28, 1989.
Niskier, Arnaldo. 1970. *Ciência e Tecnologia para o Desenvolvimento*. Rio de Janeiro: Editora Bruguera.
Nussbaum, Bruce. 1990. *Good Intentions: How Big Business and the Medical Establishment Are Corrupting the Fight against AIDS*. New York: Atlantic Monthly Press.
O'Dougherty, Maureen. 1995. International Bargain Shoppers. Paper presented at the 19th meeting of the Latin American Studies Association, Washington, D.C., September.
O'Neill, John. 1990. AIDS as a Globalizing Panic. In *Theory, Culture and Society, Special Issue on Global Culture*, ed. Mike Featherstone, 329–342. London: Sage.
Oldroy, David R. 1990. Picking at/on pickering: the deconstruction of the social construction of scientific knowledge. *Social Studies of Science* 20: 638–657.
ONGs/AIDS do Brasil. 1991. Protocolo internacional de vacinas para o HIV/AIDS em países de Terceiro Mundo: Documento das ONGs/AIDS do Brasil. *Boletim Abia* 15: 6–7.
Oppenheimer, Gerald M. 1988. In the Eye of the Storm: The Epidemiological

Construction of AIDS. In *AIDS: The Burdens of History,* ed. Elizabeth Fee and Daniel Fox, 267–300. Berkeley: University of California Press.

Packard, Randall, and Paul Epstein. 1991. Epidemiologists, social scientists, and the structure of medical research in Africa. *Soc Sci Med* 33(7): 771–782.

Pan American Health Organization. 1989. *AIDS: Profile of an Epidemic*. Washington, D.C.: PAHO/WHO.

Paiva, Vera, ed. 1992. *Em Tempos de AIDS*. São Paulo: Summus.

Panem, Sandra. 1988. *The AIDS Bureaucracy*. Cambridge, Mass.: Harvard University Press.

Pannier, Frederico. 1979. Science in Latin America. In *Science and the Factors of Inequality,* ed. Charles Morazé, 226–233. Paris: UNESCO.

Panos Dossier. 1989. *AIDS and the Third World*. Alexandria, Va.: The Panos Institute, and Santa Cruz, Calif.: New Society Publishers.

Panos Dossier. 1990. *The Third Epidemic: Repercussions of the Fear of AIDS*. London: Panos Institute.

Parker, Richard G. 1987. Acquired immunodeficiency syndrome in urban Brazil. *Medical Anthropology Quarterly* 1(2): 155–175.

Parker, Richard G. 1988. Sexual Culture and AIDS Education in Urban Brazil. In *AIDS 1988: AAAS. Symposia Papers,* ed. Ruth Kulstad, 169–173. Washington, D.C.: American Association for the Advancement of Science.

Parker, Richard G. 1989a. Respostas à AIDS no Brasil. *Boletim Abia* 6: 10–11.

Parker, Richard G. 1989b. Bodies and pleasures: on the construction of erotic meanings in contemporary Brazil. *Anthropology and Humanism Quarterly* 14(2): 58–64.

Parker, Richard G. 1991a. *Bodies, Passions and Pleasures: Sexual Culture in Contemporary Brazil*. Boston: Beacon Press.

Parker, Richard G. 1991b. *Corpos, Prazeres e Paixões: A Cultura Sexual no Brasil Contemporâneo.* São Paulo.

Parker, Richard G. 1992. Sexual Diversity, Cultural Analysis, and AIDS Education in Brazil. In *The Time of AIDS: Social Analysis, Theory and Method,* ed. Gilbert Herdt and Shirley Lindenbaum, 225–242. Newbury Park, Calif.: Sage.

Parker, Richard G. 1994a. Public Policy, Political Activism, and AIDS in Brazil. In *Global AIDS Policy,* ed. Douglas Feldman, 28–46. Westport, Conn.: Bergin and Garvey.

Parker, Richard G. 1994b. Projeto homossexualidades: ano e meio de trabalho. *Boletim Abia* l: 1–2.

Parker, R. G., and M. Carballo. 1990. Qualitative research on homosexual and bisexual behavior relevant to HIV/AIDS. *The Journal of Sex Research* 27: 497–525.

Parker, R. G., and M. Carballo. 1991. Bisexual Behavior, HIV Transmission, and Reproductive Health. In *AIDS and Reproductive Health*, ed. L. Chen and J. Sepulveda, 109–117. New York: Plenum Press.

Parker, R. G., G. Herdt, and M. Carballo. 1991. Sexual culture, HIV transmission, and AIDS research. *The Journal of Sex Research* 28: 77–98.

Parker, Richard, and Jane Galvão, eds. 1994. *Seminário "A Epidemiologia Social da AIDS": Anais.* Rio de Janeiro: IMS–UERJ/ABIA.

Parker, Richard G., Murillo P. Mota, and Lourenço E. L. Rodrigues. 1994. Sexo entre homens: uma pesquisa sobre a consciência da AIDS e comportamento (homo)-sexual no Brasil. *Boletim Abia* especial: 4–7.

Parker, Richard, Cristiana Bastos, Jane Galvão, and José Stalin Pedrosa, eds. 1994. *A AIDS no Brasil* (1982–1992). Rio de Janeiro: ABIA/IMS–UERJ/Relume–Dumará.
Patton, Cindy. 1985. *Sex and Germs: The Politics of AIDS*. Boston: South End Press.
Patton, Cindy. 1990. *Inventing AIDS*. New York: Routledge.
Paty, Michel. 1992. L'histoire des sciences en Amérique Latine. *Pensée* 288–289: 21–45.
Peabody, John W. 1995. An organizational analysis of the World Health Organization: narrowing the gap between promise and performance. *Soc Sci Med* 40(6): 731–742.
Penley, Constance, and Andrew Ross, eds. 1991. *Technoculture*. Minneapolis: University of Minnesota Press.
People With AIDS Coalition. 1985. People with AIDS Coalition Newsline 1.
Perez, Jorge, Guillermo de la Portilla, Mariluz Rodriguez, and Juan C. de la Concepcion. 1994. Experiência Cubana na Abordagem da Infecção por HIV/AIDS. Anais do Seminário A Epidemiologia Social da AIDS. Richard Parker and Jane Galvão, eds. 50–57. Rio de Janeiro: IMS/UERJ/ABIA.
Perlongher, Nestor. 1987. *O negócio do Michê: Prostituição Viril em São Paulo*. São Paulo: Ed. Brasiliense.
Perlongher, Nestor. 1992. O desaparecimento da homossexualidade. *Boletim Abia* 16: 4–6.
Perrow, Charles, and Mauro Guillén. 1990. *The AIDS Disaster: The Failure of Organizations in New York and the Nation*. New Haven, Conn.: Yale University Press.
Physis. 1993. Epidemiologia e conhecimento médico. *Physis, Revista de Saúde Coletiva* 3(1) (special issue).
Pickering, Andrew. 1992. *Science as Practice and Culture*. Chicago: University of Chicago Press.
Pierret, Janine. 1992. Une épidémie des temps modernes. *Autrement/Mutations* 130: 17–23.
Pinch, Trevor. 1992. Opening black boxes: science, technology, and society. *Social Studies of Science* 22(3): 487–510.
Pivnick, Anitra. 1993. HIV Infection and the meaning of condoms. *Culture, Medicine and Psychiatry* 17(4): 431–452.
Plummer, Kenneth, ed. 1981. *The Making of the Modern Homosexual*. Totowa, N.J.: Barnes and Noble Books.
Polanco, Xavier. 1985. Science in the developing countries: an epistemological approach on the theory of science in context. *Quipu* 2(2): 303–318.
Polanyi, Michael. 1968. The Republic of Science: Its Political and Economic Theory. In *Criteria for Scientific Development*, ed. Edward A. Shils, 1–20. Cambridge, Mass.: MIT Press.
Pollak, Michael. 1988. *Les Homosexuals et le SIDA: Sociologie d'une Epidémie*. Paris: Metaille.
Pollak, Michael. 1992. Histoire d'une cause. *Autrement/Mutations* 130: 24–39.
Pollak, Michael. 1994. *The Second Plague of Europe: AIDS Prevention and Sexual Transmission among Men in Western Europe*. New York: Harrington Park Press.
Pollak, M., M. A. Schiltz, and L. Laurindo. 1986. Les homosexuels face à l'épidémie du SIDA. *Revue d'Epidemiologie et Santé Publique* 34: 143–153.
Pollak, Michael, with Marie-Ange Schiltz. 1987. Identité sociale et gestion d'un risque de santé. *Actes de la Recherche en Sciences Sociales* 68: 77–102.
Pollak, Michael, ed., with Geneviève Paicheler and Janine Pierret. 1992. AIDS: a

problem for sociological research, trend report, *Current Sociology/La Sociologie Contemporaine* 40(3).

Possas, Cristina. 1989. *Epidemiologia e Sociedade: Heterogeneidade Estrutural e Saúde no Brasil*. São Paulo: HUCITEC.

Prescott, Frank. 1994. *Medical Doctor Puts His Life on the Line to Prove That Sex & HIV Do Not Cause AIDS*. New York: HEAL paper.

Preston, Richard. 1994. *The Hot Zone*. New York: Random House.

Quemmel, Renato. 1994. Um projeto a quatro mãos. *Boletim Abia* 1: 3.

Quimby, Ernest. 1992. Anthropological Witnessing for African Americans: Power, Responsibility, and Choice in the Age of AIDS. In *The Time of AIDS*, ed. Gilbert Herdt and Shirley Lindenbaum, 150–185. London: Sage.

Rabinow, Paul. 1993a. *Exponential Amplification: The Invention, Development, and Standardization of the Polymerase Chain Reaction*. Cambridge: Zone Books–MIT Press.

Rabinow, Paul. 1993b. Galton's regret and DNA typing. *Culture, Medicine and Psychiatry* 17(1): 59–65.

Ramos, Sílvia. 1988. Um rosto de mulher. *Boletim Abia* 2: 6–7.

Ramos, Sílvia. 1989. Sex, drugs, AIDS e Sarney (a pior AIDS do mundo). *Boletim Abia* 6: 8–9.

Ramos, Sílvia. 1990. AIDS e sociedade civil no Brasil. Paper presented at the Seventh National Meeting of Population Studies.

Rapp, Rayna. 1993. Accounting for Amniocentesis. In *Knowledge, Power and Practice*, ed. Shirley Lindenbaum and Margaret Lock, 55–76 Berkeley: University of California Press.

Rapp, Rayna. 1994. Heredity, or Revising the Facts of Life. In *Naturalizing Power: Essays in Feminist Cultural Analysis*, ed. S. Yanagisako and C. Delaney, 69–86. New York: Routledge.

Ratner, Mitchell S., ed. 1993. Crack pipe as pimp: an ethnographic investigation of sex-for-crack. *Exchanges*. New York: Lexington Books.

Rede Brasileira de Solidariedade (ONGs/AIDS). 1989. *Boletim Abia* 10–11.

Rede Brasileira de Solidariedade (ONGs/AIDS). 1989a. *Relatório do II Encontro da Rede Brasileira de Solidariedade (ONGs/AIDS)*. Porto Alegre, mim.

Rede Brasileira de Solidariedade (ONGs/AIDS). 1989b. Proposta de carta de princípios. *Boletim Abia* 9: 3–4.

Rede Brasileira de Solidariedade (ONGs/AIDS). 1989c. Declaração dos direitos da pessoa portadora do vírus da AIDS. *Boletim Abia* 9: 4.

Rede Brasileira de Solidariedade (ONGs/AIDS). 1990. Princípios Constitucionais Da Rede Brasileira De Solidariedade (ONGs/AIDS)—Documento Aprovado No II Encontro Da Rede Brasileira De Solidariedade (ONGs/AIDS), Porto Alegre.

Reid, Inez Smith. 1975. Science, Politics, and Race. *Signs* 1(2). Reprinted in *Sex and Scientific Inquiry*, ed. Sandra Harding and Jean F. O'Barr, 99–124. Chicago: University of Chicago Press.

Reif, F. 1961. The competitive world of the pure scientist. *Science* 134 (3494): 1957–1962.

Reily, Charles, ed. 1995. *New Paths to Democratic Development in Latin America: The Rise of NGO–Municipal Collaboration*. Boulder, Colo.: Lynne Rienner.

Reinfeld, Rev. Margaret R. 1994. The Gay Men's Health Crisis: A Model for Community Intervention. In *AIDS Prevention and Services: Community Based Research*, ed. Johannes P. Van Vugt, 179–198. Westport, Conn.: Bergin and Garvey.

Restivo, Sal. 1995. The Theory Landscape in Science Studies: Sociological Traditions.

In *Handbook of Science and Technology Studies,* ed. Sheila Jasanoff et al., 95–110. London: Sage.

Risse, Guenter B. 1988. Epidemics and History: Ecological Perspectives and Social Responses. In *AIDS: The Burdens of History,* ed. Elizabeth Fee and Daniel Fox, 33–66. Berkeley: University of California Press.

Robertson, Roland. 1992. *Globalization: Social Theory and Global Culture.* London: Sage.

Rogers, David E. 1986. Where have we been? Where are we going? *Daedalus* 115(2): 209–229.

Root-Bernstein, Robert. 1993. *Rethinking AIDS: The Tragic Cost of Premature Consensus.* New York: Free Press.

Rosenberg, Charles E. 1987. *The Cholera Years: The United States in 1832, 1849, and 1866.* Chicago: University of Chicago Press.

Rosenberg, Charles E. 1989. What Is an Epidemic? AIDS in Historical Perspective. In *Living with AIDS,* ed. Stephen Graubard, 1–17. Cambridge, Mass.: MIT Press.

Ross, Andrew. 1991. *Strange Weather: Culture, Science and Technology in the Age of Limits.* London: Verso.

Ross, Andrew, ed. 1996. *Science Wars.* Durham, N.C.: Duke University Press.

Rostow, W. W. 1960. *The Stages of Economic Growth: A Non-Communist Manifesto.* Cambridge: Cambridge University Press.

Rothermel, Timothy S. 1988. The Impact of AIDS in Development Programmes. In *The Global Impact of AIDS,* ed. A. F. Fleming et al., 161–165. New York: Alan R. Liss.

Rouse, Joseph. 1987. *Knowledge and Power: Toward a Political Philosophy of Science.* Ithaca, N.Y.: Cornell University Press.

Rowe, Lori. 1994. Training Tribal Leaders in Thailand. *Exchanges* 1994(2): 3–6.

Rubinstein, Robert A., Charles D. Laughlin Jr., and John McManus. 1984. *Science as Cognitive Process: Toward an Empirical Philosophy of Science.* Philadelphia: University of Pennsylvania Press.

Sabatier, Renée. 1988. *Blaming Others: Prejudice, Race and Worldwide AIDS.* Panos Institute/Norwegian Red Cross, and Philadelphia, Penn.: New Society Publishers.

Sabroza, Paulo Chagastelles, et al. 1995. Doenças transmissíveis: ainda um desafio. In *Os Muitos Brasis,* ed. Maria Cecília Minayo, 177–244.

Sagasti, Francisco. 1971. Underdevelopment, science and technology: the point of view of the underdeveloped countries. *Science Studies* 3(1): 47–59.

Salomon, Jean-Jacques, Francisco R. Sagasti, and Céline Sachs-Jeantet, eds. 1994. *The Uncertain Quest: Science, Technology, and Development.* Tokyo: United Nations University Press.

Sardar, Ziauddin, ed. 1988. *The Revenge of Athena: Science, Exploitation, and the Third World.* New York: Mansell.

Schechter, M., et al. 1994. Co-infection with human t-cell lymphotropic virus type I and HIV in Brazil: impact markers of HIV disease progression. *JAMA* 271(5): 353–357.

Schecter, Stephen. 1990. *The AIDS Notebooks.* Albany, N.Y.: SUNY Press.

Scheper-Hughes, Nancy. 1994. AIDS and the social body: an essay. *Soc Sci Med* 39(7): 991–1003.

Schiller, Nina Glick. 1993. The Invisible Women: Care-giving and the Construction of AIDS Health Services. *Culture, Medicine and Psychiatry* 17(4): 487–512.

Schiller, Nina Glick, Stephen Crystal, and Denver Lewellen. 1994. Risky business: the cultural construction of AIDS risk groups. *Soc Sci Med* 38(10): 1337–1346.

Schoepf, Brooke Grundfest. 1991. Ethical, methodological and political issues of AIDS research in Central Africa. *Soc Sci Med* 33(7): 749–764.

Schoepf, Brooke Grundfest. 1992a. Women at Risk: Cases Studies from Zaire. In *The Time of AIDS,* ed. Gilberdt Herdt and Shirley Lindenbam, 259–286. London: Sage.

Schoepf, Brooke Grundfest. 1992b. Political Economy, Sex and Cultural Logics: A View from Zaire. Paper presented at the conference Culture, Sexual Behavior and AIDS, Amsterdam, July 24–26.

Schopper, Doris. 1990. Research on AIDS interventions in developing countries: state of the art. *Soc Sci Med* 30(12): 1265–1272.

Schwartzman, Simon. 1978. "Brain drain": Pesquisa multinacional? In *A Aventura Sociológica: Objectividade, Paixão, Improviso e Método na Pesquisa Social,* ed. Edson Oliveira Nunes. Rio de Janeiro: Zahar Editores.

Schwartzman, Simon. 1979. Formação da comunidade científica no Brasil. São Paulo: Editora Nacional.

Schwartzman, Simon. 1981. *Ciência, Universidade, e Ideologia: A Política do Conhecimento.* Rio de Janeiro: Zahar.

Schwartzman, Simon, ed. 1982. *Universidades e Instituições Científicas no Rio de Janeiro.* Brasília: CNPq.

Schwarzstein, Jacques. 1992. Governo e ONGs/AIDS: consenso à vista? Foi de repente, como tudo acontece. *Boletim Abia* 17: 3–4.

Schwarzstein, Jacques. 1993a. Projeto do Banco Mundial: uma virada na história das epidemias de AIDS e DSTs no Brasil? *Boletim Abia* 19: 2–9.

Schwarzstein, Jacques. 1993b. Testagem de vacina anti-HIV no Brasil: Ciência, ficção ou ficção científica? *Boletim Abia* 20: 11–12.

Science et Vie. 1992. Dossier SIDA. *Science et Vie* 179 (special issue).

Scientific American. 1988. What science knows about AIDS. *Science* 259(4) (special issue).

Shapin, Steven. 1989. Review of *The Rational and the Social,* by James Robert Brown. *Philosophy of Science* 59(4): 712–713.

Shapin, Steven. 1992. Discipline and bounding: history and sociology of science as seen through the externalism–internalism debates. *History of Science* 30(4): 333–369.

Sheriff, Robin E. 1995. The "Seventh Son": Double-Consciousness and Discourses on Race in Rio de Janeiro. Paper presented at the 94th Annual Meeting of the American Anthropological Association, Washington, D.C., November.

Shilts, Randy. 1987. *And the Band Played On: Politics, People, and the AIDS Epidemic.* New York: St. Martin's Press.

Shrum, Wesley, and Yehouda Shevan. 1995. Science and Technology in Less Developed Countries. In *Handbook of Science and Technology Studies,* ed. Sheila Jasanoff et al., 627–651. Thousand Oaks, Calif.: Sage Publications.

Siegel, Larry, ed. 1988. *AIDS and Substance Abuse.* New York: Harrington Park Press.

Sigerist, Henry E. 1962. *Civilization and Disease.* Chicago: University of Chicago Press.

Signorile, Michelangelo. 1993. *Queer in America: Sex, the Media, and the Closets of Power.* New York: Doubleday.

Silva, Luís Pereira da. 1980. *La biologie moderne et la santé de l'homme du tiers monde.* In *Le Point Critique. Reports for the U.N. Conference in Science and Technology for Development,* ed. Charles Morazé, 201–218. Paris: Presses Universitaires de France.

Singer, Merrill. 1994a. The politics of AIDS. *Soc Sci Med* 38(10): 1321–1324.

Singer, Merrill. 1994b. AIDS and health crisis of U.S. urban poor: the perspective of critical medical anthropology. *Soc Sci Med* 39(7): 931–948.

Singer, Merrill, William Gonzalez, Evelyn Vega, Yvonne Centeno, and Lani Davidson. 1994. Implementing a Community Based AIDS Prevention Program for Ethnic Minorities: The Comunidad y Responsibilidad Project. In *AIDS Prevention and Services: Community Based Research,* ed. Johannes Van Vugt, 59–92. Westport, Conn.: Bergin and Garvey.

Skidmore. Thomas E. 1992. EUA bi-racial vs. Brasil multirracial: O contraste ainda é válido? *Novos Estudos CEBRAP* 34: 49–62.

Slack, Paul. 1985. *The Impact of Plague in Tudor and Stuart England.* London: Routledge and Kegan Paul.

Sobo, E. S. 1993. Inner-city women and AIDS: the psycho-social benefits of unsafe sex. *Culture Medicine and Psychiatry* 17(4): 455–485.

Social Science and Medicine. 1991. Social science perspectives on HIV in the United States. *Soc Sci Med* 33(1) (special issue).

Solano Vianna, Nelson. 1992. AIDS no Brasil: Avaliando o passado e planejando o futuro. Paper delivered in the seminar "AIDS e Ativism Político," Social Medicine Institute, State University of Rio de Janeiro, May 11–13, 1992.

Solano Vianna, Nelson. 1993. A solidariedade é uma grande empresa: Um projeto da ABIA para controle da AIDS no local de trabalho. *Boletim Abia* 21: 8.

Sonnabend, Joseph A., Stephen S. Witkin, and David T. Purtilo. 1984. Acquired immune deficiency syndrome: an explanation for its occurrence among homosexual men. In *The Acquired Immune Deficiency Syndrome and Infections of Homosexual Men,* ed. Pearl Ma and Donald Armstrong, 409–425. New York: Yorke Medical Books.

Sontag, Susan. 1979. *Illness as a Metaphor*. New York: Vintage.

Sontag, Susan. 1989. *AIDS as a Metaphor.* New York: Farrar, Straus and Giroux.

Souza, Alícia Regina Navarro Dias de. 1988. *A Reflexão do saber sobre a Impotência: SIDA/AIDS, Uma Experiência Em Psicologia Médica*. M.A. dissertation in Psychiatry, UFRJ, Rio de Janeiro.

Spheric. 1994. Is AIDS a hoax? *Spheric* 7: 1. Feature article.

Stall, Ron, and James Wiley. 1988. A comparison of drug and alcohol use habits of heterosexual and homosexual men. *Drug and Alcohol Dependence* 22(1): 63–74.

Stein, Howard. 1990. *American Medicine as Culture*. Boulder, Colo.: Westview Press.

Stepan, Nancy. 1976. *Beginnings of Brazilian Science: Oswaldo Cruz, Medical Research and Policy, 1890–1920*. New York: Science History Publications.

Sterk-Elifson, Claire. 1993. Risk Perception and Behavior Change among Female Drug Users. *Practicing Anthropology* 15(4): 62–65.

Strathern, Marilyn. 1992. *Reproducing the Future: Anthropology, Kinship and the New Reproductive Technologies*. New York: Routledge.

Studer, Kenneth, and Daryl E. Chubin. 1980. *The Cancer Mission: Social Contexts in Biomedical Research.* London: Sage.

Sullivan, Andrew. 1996. When plagues end: notes on the twilight of an epidemic. *New York Times Magazine,* November 10, 52–76.

Sutmoller, Frits, Cristiana Bastos, Terezinha Penna, and Claudia Monteiro. 1994. Interactions between NGOs and a Research Center for the Implementation of an AIDS Prevention Program Associated to HIV Vaccine Development. Paper presented at the Tenth International Conference on AIDS, Yokohama, August.

Tauber, Alfred I. 1994. *The Immune Self: Theory or Metaphor?* Cambridge: Cambridge University Press.

Tema/Radis. 1988. *Sangue.* Rio de Janeiro: Fiocruz (special issue).

Terto, Veriano, Jr. 1988. *No Escurinho Do Cinema: Orgia Nas Tardes Cariocas.* M.A. dissertation in Psychology, Pontificia Universidade Católico, Rio de Janeiro.

Terto, Veriano, Jr. 1993. Homossexualidades: um projeto da ABIA para a prevenção de AIDS entre homens que fazem sexo com homens. *Boletim Abia* 21: 5–7.

Thiandière, Claude, ed. 1992. L'homme contaminé: La tourmente du SIDA. *Autrement/Mutations* 130 (special issue).

Thielen, Eduardo Vilela, et al. 1991. *Science Heading for the Backwoods: Images of the Expeditions Conducted by the Oswaldo Cruz Institute Scientists to the Brazilian Hinterland, 1911/1913* [English translation by Anthony Zinesky]. Rio de Janeiro: Oswaldo Cruz Foundation, Casa de Oswaldo Cruz.

Thomas, Lewis. 1974. *The Lives of a Cell.* New York: Viking.

Thomas, Lewis. 1977. The Science and Technology of Medicine. In *Doing Better and Feeling Worse,* ed. John Knowles, 35–46. New York: Norton.

Torino, E., and A. Creese, eds. 1990. *Achieving Health for All by the Year 2000: Midway Report of Country Experiences.* Geneva: WHO.

Traweek, Sharon. 1988. *Beamtimes and Lifetimes: The World of High Energy Physicists.* Cambridge, Mass.: Harvard University Press.

Traweek, Sharon. 1993. An Introduction to Cultural and Social Studies of Science and Technologies. *Culture Medicine and Psychiatry* 17(1): 3–25.

Threichler, Paula. 1987. AIDS, Homophobia and Biomedical Discourse: An Epidemic of Signification. *October* 43: 31–70.

Threichler, Paula. 1988. AIDS, Gender, and Biomedical Discourse: Current Contests for Meaning. In *AIDS: The Burdens of History,* ed. Elizabeth Fee and Daniel Fox, 190–266. Berkeley: University of California Press.

Threichler, Paula. 1989. AIDS and HIV Infection in the Third World: A First World Chronicle. In *Remaking History,* ed. Barbara Kruger and Phil Mariani, 31–86. Seattle: Bay Press.

Threichler, Paula. 1991. How to Have Theory in an Epidemic: The Evolution of AIDS Treatment Activism. In *Technoculture,* ed. Constance Penley and Andrew Ross, 57–106. Minneapolis: University of Minnesota Press.

Threichler, Paula. 1992. AIDS, HIV, and the Cultural Construction of Reality. In *The Time of AIDS,* ed. Gilbert Herdt and Shirley Lindenbaum, 65–98. London: Sage.

Tuchel, Tammy L., and Douglas A. Feldman. 1993. A preliminary ethnography of HIV positive women in Dade County jails. *Practicing Anthropology* 15(4): 52–55.

Turnbull, David. 1991. Local knowledge and "absolute standards": a reply to Daly. *Social Studies of Science* 21: 571–573.

Turnbull, David, Max Charlesworth, Lyndsay Farral, and Terry Stokes. 1989. *Life among Scientists: An Anthropology of an Australian Scientific Community.* Melbourne: Oxford University Press.

Twine, Francine Winddance. 1995. Towards an Engendered Analysis of Encounters with Racism: The Case of Upwardly Mobile Afro-Brazilians. Paper presented at the 94th Annual Meeting of the American Anthropological Association, Washington, D.C., November.

Twine, Francine Winddance. 1997. *Racism in a Racial Democracy: The Maintenance of a White Supremacy in Brazil.* New Brunswick, N.J.: Rutgers University Press.

Tylor, Edward Burnett. 1903. *Primitive Culture: Researches into the Development of Mythology, Philosophy, Religion, Language and Custom.* 4th ed., rev. London: Murray.

Valinotto, Tereza Christina. 1990. *As ONGs/AIDS: Movimento social no Brasil.* Field Report, mim.

Valinotto, Tereza Christina. 1991. *A Construção da Solidariedade: Análise das Respostas Coletivas à AIDS.* M.A. dissertation, National School of Public Health, FIOCRUZ, Rio de Janeiro.

Valle, Carlos Guilherme O. 1994. *Pessoas Vivendo com AIDS: Política e Identidade nos anos 90.* Research proposal, Anthropology Department, Museu Nacional, UFRJ.

Valle, Victor Vincent, and Luiz Werneck da Silva. 1981. *Ciência e tecnologia no Brasil: História e ideologia (1949–1976).* Brasília: CNPq.

Van Vugt, Johannes P. 1994. The Effectiveness of Community Based Organizations in the Medical Social Sciences: A Case Study of a Gay Community's Response to the AIDS Crisis. In *AIDS Prevention Services: Community Based Research,* ed. Johannes P. Van Vugt, 13–34. Westport, Conn.: Bergin and Garvey.

Van Vugt, Johannes P., ed. 1994. *AIDS Prevention Services: Community Based Research.* Westport, Conn.: Bergin and Garvey.

Varela, Francisco. 1979. *Principles of Biological Autonomy.* New York: Elsevier North Holland.

Varela, Francisco. 1988. Structural Coupling and the Origin of Meaning in a Simple Cellular Automation. In *The Semiotics of Cellular Communication in the Immune System,* ed. E. E. Sercarz et al., 151–161. Berlin: Springer-Verlag.

Varela, Francisco, and António Coutinho. 1991. Second generation immune networks. *Immunology Today* 12: 159–166.

Varela, Francisco, et al. 1988. Cognitive Networks: Immune, Neural, and Otherwise. In *Theoretical Immunology, Part 2,* ed. A. S. Perelson, 359–375. Reading, Mass.: Addison–Wesley.

Vaz, Nelson, and Francisco Varela. 1978. Self and nonself: an organism-centered approach to immunology. *Medical Hypotheses* 4: 231–267.

Vaz, Nelson, with Ana Maria C. de Faria. 1992. *Guia Incompleto de Imunobiologia.* Belo Horizonte, mim.

Vessuri, H. 1987. The social study of science in Latin America. *Social Studies of Science.* 17: 519–554.

Vogel, Morris, and Charles Rosenberg, eds. 1979. *The Therapeutic Revolution.* Philadelphia: University of Pennsylvania Press.

Von Giziky, Rainald. 1987. Cooperation between Medical Researchers and a Self-Help Movement: The Case of the German Retinitis Pigmentosa Society. In *The Social Direction of the Public Sciences,* ed. Stuart Blume et al., 75–88. Dordrecht: D. Reidel.

Wallerstein, Immanuel. 1974–1980. *The Modern World-System.* New York: Academic Press.

Ward, Martha C. 1993. A different disease: HIV/AIDS and health care for women in poverty. *Culture Medicine and Psychiatry* 17(4): 413–430.

Watney, Simon. 1990. Safer Sex as Community Practice. In *AIDS: Individual, Cultural, and Policy Dimensions,* ed. Peter Aggleton, Peter Davies, and Graham Hart, 19–30. London: Falmer Press.

Watney, Simon. 1994. *Practices of Freedom: Selected Writings on HIV/AIDS.* London: Rivers Oram Press.

Weber, Max. 1963. *The Sociology of Religion.* Boston: Beacon Press.

Weil, Diana E., et al. 1990. *The Impact of Development Policies on Health: A Review of the Literature*. Geneva: WHO.

Weindling, Paul. 1992. From Infections to Chronic Disease: Changing Patterns of Sickness in the Nineteenth and Twentieth Centuries. In *Medicine and Society*, ed. Andrew Wear, 303–316. Cambridge: Cambridge University Press.

Whiten, Frederick. 1995 [1979]. Os Entendidos: Gay Life in São Paulo in the Late 1970s. In *Latin American Male Homosexualities*, ed. Stephen O. Murray, 231–240. Albuquerque: University of New Mexico Press.

WHO (World Health Organization). 1962. *Fighting Disease*. Dobbs Ferry, N.Y.: Oceana Publications.

WHO (World Health Organization). 1975. *Health by the People*. Geneva: WHO.

WHO (World Health Organization). 1976. *Introducing WHO*. Geneva: WHO.

WHO (World Health Organization). 1979. *Formulating Strategies for Health for All in the Year 2000*. Geneva: WHO.

WHO (World Health Organization). 1980. *Sixth Report on the World Health Situation 1973–1977*. Geneva: WHO.

WHO (World Health Organization). 1981. *Global Strategies for Health for All by the Year 2000*. Geneva: WHO.

WHO (World Health Organization). 1985a. AIDS: where do we go from here? *WHO Chronicle* 39(3): 98–103.

WHO (World Health Organization). 1985b. AIDS: the search for clues. *WHO Chronicle* 39(6): 207–211.

WHO (World Health Organization). 1986a. *Evaluation of the Strategy for Health for All by the Year 2000. Seventh Report on the World Health Situation*, vol. 1: *Global Review*. Geneva: WHO.

WHO (World Health Organization). 1986b. WHO activities for AIDS prevention and control. *WHO Chronicle* 40(2): 52–53.

WHO (World Health Organization). 1987. *Tropical Disease Research: A Global Partnership*. Geneva: WHO.

WHO (World Health Organization). 1988a. *Four Decades of Achievement: Highlights of the Work of WHO*. Geneva: WHO.

WHO (World Health Organization). 1988b. AIDS Prevention and Control. In *Invited presentations and papers from the World Summit of Ministers of Health and Programmes for AIDS Prevention*. Jointly organized by the World Health Organization and the United Kingdom Government.

WHO (World Health Organization). 1988c. *Guidelines for the Development of a National AIDS Prevention Programme*. Geneva: WHO.

WHO (World Health Organization). 1989a. *Monitoring of National AIDS Prevention and Control Programmes: Guiding Principles*. Geneva: WHO.

WHO (World Health Organization). 1989b. *Guide to Planning Health Promotion for AIDS Prevention and Control*. Geneva: WHO.

WHO (World Health Organization). 1989c. *Non-Governmental Organizations and the Global AIDS Strategy. World Health Assembly, Resolution #42, 34*. Geneva: WHO.

WHO (World Health Organization). 1992. *The Global AIDS Strategy*. Geneva: WHO.

WHO (World Health Organization). 1993a. *Implementation of the Global Strategy for Health for All by the Year 2000: Second Evaluation. Eighth Report on the World Health Situation*, vol. 3: *Region of the Americas*. Geneva: WHO.

WHO (World Health Organization). 1993b. *The Urban Health Crisis: Strategies for Health for All in the Face of Rapid Urbanization*. Geneva: WHO.

WHO (World Health Organization). 1993c. *Global Programme on AIDS—1991 Progress Report.* Geneva: WHO.

WHO (World Health Organization). 1994a. *Ninth General Programme of Work Covering the Period 1996–2001.* Geneva: WHO.

WHO (World Health Organization). 1994b. *AIDS: Images of the Epidemic.* Geneva: WHO.

Wojnarowicz, David. 1991. *Close to the Knives: A Memoir of Disintegration.* New York: Vintage Books.

Wolfe, Maxine. 1994. The AIDS Coalition to Unleash Power (ACT UP): A Direct Model of Community Research for AIDS Prevention. In *AIDS Prevention Services: Community Based Research,* ed. Johannes P. Van Vugt, 217–247. Westport, Conn.: Bergin and Garvey.

Worth, Dooley. 1990. Minority Women and AIDS: Culture, Race, and Gender. In *Culture and AIDS,* ed. Douglas A. Feldman, 111–135. New York: Praeger.

Wright, Peter, and Andrew Treacher. 1982. *The Problem of Medical Knowledge: Examining the Social Construction of Medicine.* Edinburgh: Edinburgh University Press.

Index

CRISTIANA BASTOS is an anthropologist and senior researcher at the Social Sciences Institute, University of Lisbon. She is the author of *Os Montes do Nordeste Algarvio* (1993) and coeditor (with Richard Parker, Jane Galvão, and José Stalin Pedrosa) of *A AIDS no Brasil (1892–1992)* (1994).